DELAY FAULT TESTING
FOR
VLSI CIRCUITS

FRONTIERS IN ELECTRONIC TESTING

Consulting Editor
Vishwani D. Agrawal

Books in the series:

Research Perspectives and Case Studies in System Test and Diagnosis
J.W. Sheppard, W.R. Simpson
ISBN: 0-7923-8263-3

Formal Equivalence Checking and Design Debugging
S.-Y. Huang, K.-T. Cheng
ISBN: 0-7923-8184-X

On-Line Testing for VLSI
M. Nicolaidis, Y. Zorian
ISBN: 0-7923-8132-7

Defect Oriented Testing for CMOS Analog and Digital Circuits
M. Sachdev
ISBN: 0-7923-8083-5

Reasoning in Boolean Networks: Logic Synthesis and Verification Using Testing Techniques
W. Kunz, D. Stoffel
ISBN: 0-7923-9921-8

Introduction to I_{DDQ} Testing
S. Chakravarty, P.J. Thadikaran
ISBN: 0-7923-9945-5

Multi-Chip Module Test Strategies
Y. Zorian
ISBN: 0-7923-9920-X

Testing and Testable Design of High-Density Random-Access Memories
P. Mazumder, K. Chakraborty
ISBN: 0-7923-9782-7

From Contamination to Defects, Faults and Yield Loss
J.B. Khare, W. Maly
ISBN: 0-7923-9714-2

Efficient Branch and Bound Search with Applications to Computer-Aided Design
X.Chen, M.L. Bushnell
ISBN: 0-7923-9673-1

Testability Concepts for Digital ICs: The Macro Test Approach
F.P.M. Beenker, R.G. Bennetts, A.P. Thijssen
ISBN: 0-7923-9658-8

Economics of Electronic Design, Manufacture and Test
M. Abadir, A.P. Ambler
ISBN: 0-7923-9471-2

I_{DDQ} Testing of VLSI Circuits
R. Gulati, C. Hawkins
ISBN: 0-7923-9315-5

DELAY FAULT TESTING FOR VLSI CIRCUITS

Angela Krstić
University of California/Santa Barbara

Kwang-Ting (Tim) Cheng
University of California/Santa Barbara

KLUWER ACADEMIC PUBLISHERS
Boston / Dordrecht / London

Distributors for North, Central and South America:
Kluwer Academic Publishers
101 Philip Drive
Assinippi Park
Norwell, Massachusetts 02061 USA
Telephone (781) 871-6600
Fax (781) 871-6528
E-Mail <kluwer@wkap.com>

Distributors for all other countries:
Kluwer Academic Publishers Group
Distribution Centre
Post Office Box 322
3300 AH Dordrecht, THE NETHERLANDS
Telephone 31 78 6392 392
Fax 31 78 6546 474
E-Mail <orderdept@wkap.nl>

 Electronic Services <http://www.wkap.nl>

Library of Congress Cataloging-in-Publication Data

A C.I.P. Catalogue record for this book is available
from the Library of Congress.

Copyright © 1998 by Kluwer Academic Publishers.

All rights reserved. No part of this publication may be reproduced, stored in a retrieval system or transmitted in any form or by any means, mechanical, photo-copying, recording, or otherwise, without the prior written permission of the publisher, Kluwer Academic Publishers, 101 Philip Drive, Assinippi Park, Norwell, Massachusetts 02061

Printed on acid-free paper.

Printed in the United States of America

CONTENTS

FOREWORD ix
PREFACE xi

1. INTRODUCTION 1
   1.1 A Problem of Interest 1
   1.2 Overview of the book 3
2. TEST APPLICATION SCHEMES FOR TESTING DELAY DEFECTS 7
   2.1 Combinational Circuits 8
   2.2 Sequential Circuits 8
       2.2.1 Enhanced scan testing 10
       2.2.2 Standard scan testing 10
       2.2.3 Slow-fast-slow clock testing 12
       2.2.4 At-speed testing 13
   2.3 Testing High Performance Circuits Using Slower Testers 14
       2.3.1 Slow-fast-slow testing strategy on slow testers 16
       2.3.2 At-speed testing strategy on slow testers 19
   2.4 Summary 22

3. DELAY FAULT MODELS — 23
- 3.1 Transition Fault Model — 23
- 3.2 Gate Delay Fault Model — 27
- 3.3 Line Delay Fault Model — 28
- 3.4 Path Delay Fault Model — 28
- 3.5 Segment Delay Fault Model — 29
- 3.6 Summary — 30

4. CASE STUDIES ON DELAY TESTING — 33
- 4.1 Summary — 44

5. PATH DELAY FAULT CLASSIFICATION — 45
- 5.1 Sensitization Criteria — 46
  - 5.1.1 Single-path sensitizable path delay faults — 47
  - 5.1.2 Robust testable path delay faults — 48
  - 5.1.3 Non-robust testable path delay faults — 50
  - 5.1.4 Validatable non-robust testable path delay faults — 52
  - 5.1.5 Functional sensitizable path delay faults — 53
- 5.2 Path Delay Faults that do Not Need Testing — 55
  - 5.2.1 Functional irredundant vs. functional redundant path delay faults — 56
  - 5.2.2 Robust vs. robust dependent path delay faults — 59
  - 5.2.3 Path classification based on input sort heuristic — 61
  - 5.2.4 Path classification based on single stuck-at fault tests — 62
  - 5.2.5 Primitive vs. non-primitive path delay faults — 63
- 5.3 Multiple Path Delay Faults and Primitive Faults — 64
- 5.4 Path Delay Fault Classification for Sequential Circuits — 66
  - 5.4.1 Sequential PDFC — 68
  - 5.4.2 Untestable segment faults — 70
  - 5.4.3 Algorithm for identifying testable PDFC for sequential circuits — 73
- 5.5 Summary — 75

6. DELAY FAULT SIMULATION — 77
- 6.1 Transition Fault Simulation — 78
  - 6.1.1 Simulating transition faults in sequential circuits — 78
- 6.2 Gate delay fault simulation — 85
- 6.3 Path Delay Fault Simulation — 88

	6.3.1	Enumerative methods for estimating path delay fault coverage	89
	6.3.2	Non-enumerative methods for estimating path delay fault coverage	92
6.4	Segment Delay Fault Simulation		98
6.5	Summary		99

7. TEST GENERATION FOR PATH DELAY FAULTS   101

- 7.1 Robust Tests   102
- 7.2 High Quality Non-Robust Tests   104
  - 7.2.1 Algorithm for generating non-robust tests with high robustness   107
- 7.3 Validatable Non-Robust Tests   112
- 7.4 High Quality Functional Sensitizable Tests   113
  - 7.4.1 Algorithm for generating high quality functional sensitizable tests   115
- 7.5 Tests for Primitive Faults   118
  - 7.5.1 Co-sensitizing gates   120
  - 7.5.2 Merging gates   123
  - 7.5.3 Identifying FS paths not involved in any primitive fault   123
  - 7.5.4 Algorithm for identifying and testing primitive faults of cardinality 2   125
- 7.6 Summary   130

8. DESIGN FOR DELAY FAULT TESTABILITY   131

- 8.1 Improving The Path Delay Fault Testability by Reducing The Number of Faults   131
- 8.2 Improving The Path Delay Fault Testability by Increasing Robust Testability of Designs   143
- 8.3 Improving Path Delay Fault Testability by Increasing Primitive Delay Fault Testability   144
  - 8.3.1 Primitive faults of cardinality $k > 2$ 146
  - 8.3.2 Design for primitive delay fault testability   148
- 8.4 Summary   154

9. SYNTHESIS FOR DELAY FAULT TESTABILITY   157

- 9.1 Synthesis for Robust Delay Fault Testability   159
  - 9.1.1 Combinational and enhanced scan sequential circuits   159
  - 9.1.2 Standard scan sequential circuits   163

9.2 Synthesis for Validatable Non-Robust Testable and Delay-Verifiable Circuits 166
9.3 Summary 167

10. CONCLUSIONS AND FUTURE WORK 169

REFERENCES 173

INDEX 189

FOREWORD

In the early days of digital design, we were concerned with the logical correctness of circuits. We knew that if we slowed down the clock signal sufficiently, the circuit would function correctly. With improvements in the semiconductor process technology, our expectations on speed have soared. A frequently asked question in the last decade has been how fast can the clock run. This puts significant demands on timing analysis and delay testing.

Fueled by the above events, a tremendous growth has occurred in the research on delay testing. Recent work includes fault models, algorithms for test generation and fault simulation, and methods for design and synthesis for testability. The authors of this book, Angela Krstić and Tim Cheng, have personally contributed to this research. Now they do an even greater service to the profession by collecting the work of a large number of researchers. In addition to expounding such a great deal of information, they have delivered it with utmost clarity. To further the reader's understanding many key concepts are illustrated by simple examples.

The basic ideas of delay testing have reached a level of maturity that makes them suitable for practice. In that sense, this book is the best available guide for an engineer designing or testing VLSI systems. Techniques for path delay testing and for use of slower test equipment to test high-speed circuits are of particular interest.

The book discusses several new fault models which, coupled with techniques of synthesis for testability, provide ideal venues for further research. The bibliography is the most complete ever put together on this topic. The reader can build the necessary foundation for the future research in the area of delay testing.

The techniques described here are applicable to the clocked synchronous design, which is the dominant style used in digital systems today. However, I believe these ideas will lead to test methodologies for the future design styles, whatever they may be. Without speculating, let me quote the visionary statements Stephen Unger makes in his 1989 book, *The Essence of Logic Circuits* (Prentice-Hall). He writes, "... the speed with which synchronous systems operate is quite sensitive to *variations* in the stray delays along paths both within the combinational logic block and in the clock distribution network. As IC technology progresses and logic elements are scaled down further in size, the *relative* size of on-chip wiring delays grows. This causes skew to become relatively large. For this reason, clock periods are not falling as fast as the long-path delays. As a result, more consideration is being given to the possibility of designing systems or parts of systems that operate without clocks at speeds determined by their own internal parameters."

Irrespective of whether we continue to use the synchronous design or develop the *speed independent* style that Unger alludes to, I am certain delay test will assume greater importance in the future than it has in the past.

Murray Hill, New Jersey

Vishwani D. Agrawal
va@research.bell-labs.com

PREFACE

With the ever increasing speed of integrated circuits, violations of the performance specifications are becoming a major factor affecting the product quality level. The need for testing timing defects is further expected to grow with the current design trend of moving towards deep submicron devices. After a long period of prevailing belief that high stuck-at fault coverage is sufficient to guarantee high quality of shipped products, the industry is forced to rethink other types of testing.

Delay testing has been a topic of extensive research both in industry and in academia for more than a decade. As a result, several delay fault models and numerous testing methodologies have been proposed. This book presents a selection of existing delay testing research results. It combines introductory material with state-of-the-art techniques that address some of the current problems in delay testing.

Following the introduction of delay testing problems and discussion of the test application schemes for detecting timing defects and different fault models which constitute the first three chapters, the book primarily focuses on the path delay fault model. Chapter 4 presents results and summary of experimental data of several case studies on delay testing performed in industry or academia. The topics covered by Chapters 5 through 9 include path delay fault classification, delay fault simulation, test generation, design and synthesis for delay fault testability. The main topics are presented through a review of a number of selected published delay testing research results combined with the material from the PhD dissertation of the first author completed under the advising of the second author (Chapters 2, 5, 7 and 8). Even though the authors' intention is to provide a comprehensive study of the problem of testing delay defects in digital circuits, the list of the reviewed and referenced work represents *their* selection and is by no means complete.

The book is intended for use by CAD and test engineers, researchers, tool developers and graduate students. It requires basic background in digital testing. The book can be used as supplementary material for a graduate-level course on VLSI testing.

Acknowledgments. The authors would like to express their gratitude to Dr. Srimat Chakradhar of C&C Research Labs, NEC USA, for continuous exchange and many fruitful discussions on various delay testing problems, as well as for the technical contributions to some of the research results presented in this book. Special thanks are due to one of the true leaders in the field of VLSI testing, Dr. Vishwani Agrawal of Bell Labs, Lucent Technologies, for his thorough review of the complete manuscript and many constructive suggestions. The authors would also like to thank Srinivas Devadas, Kurt Keutzer, Irith Pomeranz, Sudakhar Reddy and Hsi-Chuan Chen, whose work on delay testing has greatly contributed to the authors' interest in this topic.

The writing of this book was supported in part by National Science Foundation under grant MIP-9409174, by California MICRO program and by grants from NEC USA, Inc.

Santa Barbara, California

Angela Krstić
Kwang-Ting (Tim) Cheng

1 INTRODUCTION

1.1 A PROBLEM OF INTEREST

The objective of delay testing is to detect timing defects and ensure that the design meets the desired performance specifications. The need for delay testing has evolved from the common problem faced by the semiconductor industry: designs that function properly at low clock frequencies fail at the rated speed. As experiments show, tests that do not specifically target delay faults have a limited success in detecting timing defects [21, 107, 108, 109, 147].

The growing need for delay testing is a result of the advances in VLSI technology and the increase of the design speed. These factors are also changing the target objectives of delay tests. In the early days, most defects affecting the performance could be detected using tests for gross delay defects [30, 94, 135, 151]. Aggressive timing requirements of high speed designs have introduced the need to test smaller timing defects and distributed faults caused by statistical process variations [98, 139]. The increase of the circuit size has resulted in fault models that can detect distributed defects localized to a certain area of the chip [64]. With the introduction of deep submicron technology, noise effects are becoming significant contributors to timing failures and they call for further adaptations of the fault models and testing strategies.

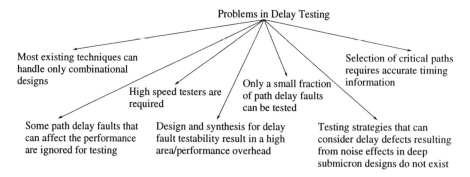

Figure 1.1. Main problems in delay testing.

Testing delay defects is a complex problem. Difficulties are related to both, the test generation and the test application process. Some of the main problems in delay testing are summarized in Figure 1.1 and described below:

- Delay defects can only be activated and observed by propagating signal transitions through the design. This requires application of vector pairs which significantly affects the delay testability of designs. Sequential circuits are especially difficult to test since an arbitrary vector pair cannot be applied to the combinational part. Most of the existing delay testing techniques are applicable only to combinational designs and cannot be easily extended to handle sequential designs. The proposed schemes for scan testing of sequential designs result either in a high area overhead and long test application time or in an unsatisfactory fault coverage. Non-scan sequential designs are often tested using variable clock speeds that complicate the test application process.

- Testing delay defects in high speed circuits requires the availability of high speed testers. However, due to the high cost, testers in the test facilities are usually several times slower than the new designs that need to be tested on them.

- Small distributed delay defects can best be modeled using the path delay fault model. However, practical designs have a very large number of paths and only a small fraction of them can be tested. Selection of paths for testing is especially difficult in performance optimized designs because they often have a large number of paths with long propagation delays.

- Selection of critical paths for testing requires accurate timing information which is not easily available. Deep submicron process introduces new difficulties into the critical path selection because noise factors such as power net noise, ground bounce and crosstalk can significantly affect the signal delays and some paths can be more sensitive to these effects than others.

- Many paths that can affect the performance of the circuit cannot be sensitized such that the detection of faults on them is independent of the delays outside the target path. Ignoring these paths contributes to a low path delay fault coverage. On the other hand, generating tests for these faults is extremely difficult because it requires considering sets of paths. It can also result in a very large test set.

- Design and synthesis for testability techniques for improving the delay fault testability usually result in a high area/performance overhead and/or large number of extra primary inputs.

- There do not exist test generation strategies that can take into account signal speedups/slowdowns resulting from noise in deep submicron devices.

This book describes some of the latest research results addressing delay testing problems. To accommodate readers not familiar with delay testing, the book provides necessary background and motivations for testing timing defects. A brief overview of the existing delay testing techniques is presented. Detailed algorithms are given for selected solutions. Most of the described techniques are illustrated through examples and summarized using flowcharts.

1.2 OVERVIEW OF THE BOOK

The book covers several delay testing topics. Chapters 2, 3 and 4 provide the necessary background and motivations for testing timing defects while the rest of the book deals with selected delay fault testing issues such as: test generation, fault simulation and design and synthesis for delay fault testability. Chapters 2, 3, 4, 6 and 9 consider several different delay fault models while Chapters 5, 7 and 8 concentrate solely on the path delay fault model.

Chapter 2 gives an overview of the test application schemes for detecting delay defects in VLSI designs. Techniques allowing application of two-vector patterns for combinational, scan and non-scan sequential circuits are described. This chapter also describes techniques that allow detection of timing failures using testers that are slower than the devices under test.

Modeling of delay defects is addressed in Chapter 3. Characteristics of several frequently used delay fault models are given. A discussion of advantages and disadvantages of transition, gate, line, path and segment delay fault models is given.

Results of several case studies on detecting timing defects are presented in Chapter 4. These results demonstrate the need and provide the motivation for delay testing.

Chapter 5 focuses on classification of path delay faults based on sensitizing conditions. Some path delay faults cannot independently affect the performance and do not need to be tested to guarantee the performance of the design. Several different techniques for partitioning the set of faults into the set that needs to be tested and the set that need not be tested are presented and compared.

Fault simulation for delay faults is considered in Chapter 6. Different algorithms for transition, path and segment delay fault simulation are outlined. Enumerative and non-enumerative algorithms for path delay fault simulation are described.

Chapter 7 addresses test generation for path delay faults. Detailed algorithms for generating tests under different propagation conditions are described. In addition to testing single path delay faults, this chapter also provides detailed algorithms for identifying and testing multiple path delay faults that can affect the circuit performance.

Design for testability techniques for delay faults are described in Chapter 8. In addition to techniques that focus on reduction of path count, this chapter also describes methods for obtaining fully robustly testable designs as well as a technique that can guarantee the performance of the design after checking all single and double path delay faults.

Chapter 9 presents synthesis for delay fault testability techniques. These techniques result either in fully robustly testable designs or in delay-verifiable designs.

Chapter 10 discusses some of the future directions in delay testing research. It describes the problems arising with the development of deep submicron process.

An extensive bibliography is provided at the end of the text. One of the earliest references on delay testing dates back to 1980 [93]. In just two decades a large amount of work has been reported. Due to the large number of publications in this area, our bibliography contains only a selection of the work most relevant to the topics covered in this book.

2 TEST APPLICATION SCHEMES FOR TESTING DELAY DEFECTS

Unlike stuck-at fault testing, delay testing is closely tied to the test application strategy. This means that before tests for delay faults are derived it is necessary to know how these tests will be applied to the circuit. The testing strategy depends on the type of the circuit (combinational, scan, non-scan or partial scan sequential) as well as on the speed of the testing equipment. Ordinarily, testing delay defects requires that the test vectors be applied to the circuit at its intended operating speed. However, since high speed testers require huge investments, testers currently used in test facilities could be slower than the new designs that need to be tested on them. Testing high speed designs on slower testers requires special test application and test generation strategies.

The focus of this chapter is on different test application schemes for combinational and sequential circuits. Techniques used for testing scan as well as non-scan designs are described. Also, the issue of testing high speed designs using slow testers is addressed and some of the currently available solutions to this problem are described.

2.1 COMBINATIONAL CIRCUITS

To observe delay defects it is necessary to create and propagate transitions in the circuit. Creating transitions requires application of a vector pair, $V = \langle v_1, v_2 \rangle$. The first vector initializes the circuit while the second vector causes the desired transitions. The test application scheme for combinational circuits is shown in Figure 2.1. In normal operation, only one clock (system clock) is used to control the input and output latches and its period is T_c. In test mode, the input and output latches are controlled by two different clocks: the input and output clocks, respectively. The period of these clocks, T_s, is assumed to be larger than T_c. The input and output clocks are skewed by an amount equal to T_c. The first vector, v_1, is applied to primary inputs at time t_0. The second

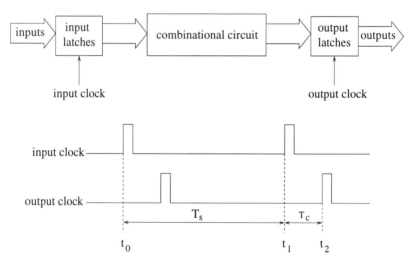

Figure 2.1. Testing scheme for combinational circuits.

vector, v_2, is applied at time t_1. Time $T_s = t_1 - t_0$ is assumed to be sufficient for all signals in the circuit to stabilize under the first vector. After the second vector is applied, the circuit is allowed to settle down only until time t_2, where $t_2 - t_1 = T_c$. At time t_2, the primary output values are observed and compared to a prestored response of a fault-free circuit to determine if there is a defect.

2.2 SEQUENTIAL CIRCUITS

Testing delay faults in sequential circuits is significantly more difficult than testing delay faults in combinational circuits. This is because appli-

TEST APPLICATION SCHEMES FOR TESTING DELAY DEFECTS 9

cation of an arbitrary vector pair is not possible to non-scan or standard scan sequential circuits. Figure 2.2(a) illustrates the Huffman model of a sequential circuit. Its operation can be represented using an iterative array of the combinational logic (shown in Figure 2.2(b)). Each copy of the combinational logic is called a **time-frame**. The present state

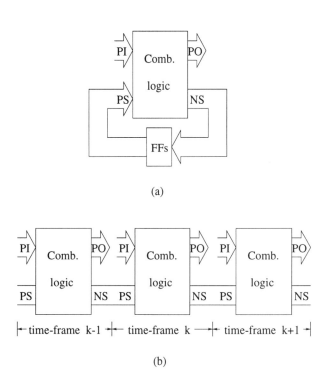

Figure 2.2. Model for sequential circuits.

(PS) values in time-frame k correspond to the next state (NS) values in time-frame $k - 1$. In the case of sequential circuits, a vector pair, $V = \langle v_1, v_2 \rangle$, can be represented as pair $V = \langle i_1 + s_1, i_2 + s_2 \rangle$, where i_1 and i_2 are the values of the primary input lines, s_1 and s_2 are values of the present state lines and symbol "+" denotes concatenation of bit vectors. Therefore, vector i_1 is required to produce s_2 as the next state of the sequential machine.

There are several commonly used testing strategies for sequential circuits: *enhanced scan*, *functional justification* and *scan shifting* for standard scan, *slow-fast-slow strategy* and *at-speed strategy* for non-scan or partial scan designs.

2.2.1 Enhanced scan testing

To be able to apply an arbitrary vector pair to the combinational portion of a sequential circuit, Dervisoglu and Strong [37] propose using memory elements that can store two bits instead of just one. Such flip-flops are called **enhanced scan flip-flops**. The design of enhanced scan flip-flops allows a pre-scan of any two-vector test. The two vectors are applied to the circuit in two consecutive clock cycles. The disadvantages of using enhanced scan flip-flops are high area overhead and long test application time.

2.2.2 Standard scan testing

Generating tests for delay faults for standard scan designs corresponds to a two time-frame sequential circuit test generation. In the first time frame, all primary inputs and present state lines are fully controllable. In the second time-frame, only the primary inputs are fully controllable. Testing schemes for standard scan have been proposed in the literature [33, 132, 133, 134]. These techniques use **functional justification** (also called **broad-side test** [134]) or **scan shifting** [33] (also called **skewed-load test** [132, 133]) to obtain the second vector. In functional justification, the second vector is derived using the capture mode and represents the set of next state values obtained after the application of the first vector. In scan shifting, the second vector is obtained by shifting the contents of the scan chain by one bit after the application of the first vector using the scan-shift mode. Figure 2.3 illustrates the functional justification and scan shifting concepts.

Cheng *et al.* [33] propose a delay test generation algorithm for standard scan designs. It is modified from a PODEM-based combinational test generator. The modifications involve a two time-frame expansion of the combinational logic of the circuit, and the use of backtracking heuristics tailored to detecting delay faults. The present state values for the second vector are generated using functional justification or scan shifting. A fault that is untestable under scan shifting might be testable under functional justification and vice versa. On an average, the test generation complexity is lower when scan shifting rather than functional justification is used.

The order of flip-flops in the scan chain cannot affect the fault coverage when functional justification is used. However, when scan shifting is applied, the order of flip-flops affects the fault coverage. To find a

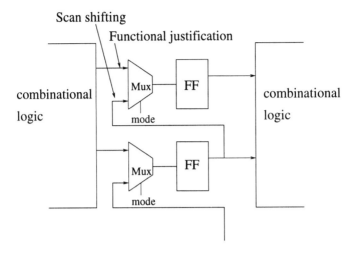

Figure 2.3. Standard scan design testing schemes.

good order of flip-flops in the scan chain, Cheng et al. [33] first apply the test generation algorithm for standard scan designs using functional justification. If the fault is not detectable using functional justification, the test generation in enhanced scan mode is tried. The test generator attempts to have as many *don't care* entries as possible in the present state lines in the second vector of the two vector sequence. Once the test pair, $\langle v_1, v_2 \rangle = \langle i_1 + s_1, i_2 + s_2 \rangle$, is generated, a set of constraints on the scan ordering is computed. These constraints, if satisfied, guarantee that s_2 can be obtained by scan shifting of v_1 in standard scan. In general, if the value of flip-flop FF_i is 0 (1) in s_1 and the value of flip-flop FF_j is 1(0) in s_2, then the constraint is that the flip-flop FF_i cannot be the immediate predecessor of flip-flop FF_j in the scan chain. If the circuit has n flip-flops, the constraints can be recorded using a quadratic matrix A of size n. Initially all entries in this matrix are set to zero. Given a test vector pair for some target fault, if flip-flop FF_i is not allowed to be the immediate predecessor of flip-flop FF_j in the scan chain, then entry A_{ij} is increased by one. The matrix A is updated after each fault in the fault list is processed, until the fault list becomes empty. The final value A_{ij} represents the number of faults that will not be detected by scan shifting if flip-flop FF_i is the predecessor of flip-flop FF_j under the vector set used to construct the matrix A. The scan ordering is determined such that most of the constraints in matrix A are satisfied. Since a delay fault can have more than one test and only one of the tests is used to construct the matrix A, the fault might be detected even if the constraints in matrix A have not been completely satisfied.

Even with an efficient order of the flip-flops in the scan chain, a certain number of faults that can be detected under enhanced scan design, cannot be detected under standard scan design. To increase the fault coverage, Cheng *at al.* [33] propose a **partial enhanced scan** design. In this design methodology, a subset of flip-flops is selected for enhanced scan. The present state lines of enhanced scan flip-flops are fully controllable in both time-frames in test generation. Given a set of faults that are testable under enhanced scan but are untestable under functional justification or scan shifting scheme with a given ordering of flip-flops, the proposed heuristic attempts to minimize the number of scan flip-flops to be enhanced in order to achieve a specified fault coverage.

2.2.3 Slow-fast-slow clock testing

Testing a fault in non-scan or partial scan sequential circuits requires a sequence of vectors. These vectors require a three phase test generation process: fault initialization, fault activation and fault propagation. Fault initialization sets the signal values to the required values for fault activation. In the fault propagation phase, the fault effect is propagated from a next state line to some primary output. Fault initialization and fault propagation require a test sequence while the fault activation requires a vector pair. The existence of delay defects in the initialization and propagation phases can interfere with activation or the observation of the fault. A common solution is to apply a **slow-fast-slow clock** testing strategy. It assumes that the vectors for initialization and propagation of the fault effect are applied at a slow speed such that the circuit can be considered delay fault-free in these test phases. In the activation phase the first vector is applied under the slow clock while the second vector is applied at the rated speed. Figure 2.4 illustrates the slow-fast-slow testing strategy.

Testing methodologies for non-scan sequential designs using the slow-fast-slow scheme have been proposed in [4, 24, 26, 38, 146]. The method proposed by Devadas [38] is based on extracting the complete or partial state transition graph. A known reset state is required. Due to the need for extracting the state transition graph, this method cannot handle large circuits. Agrawal *et al.* [4] propose inserting a logic block into the sequential circuit netlist to form a test generation model for each fault such that test generation for a delay fault becomes equivalent

TEST APPLICATION SCHEMES FOR TESTING DELAY DEFECTS 13

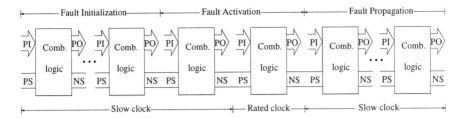

Figure 2.4. Slow-fast-slow testing strategy.

to test generation for a certain stuck-at fault. Chakraborty et al. [24] propose a delay test generator based on the iterative logic array model for sequential circuits. It considers two time-frames at a time.

Using a slow clock in fault initialization and fault propagation phases significantly simplifies the test generation for delay faults. However, the need for two clocks (slow and fast) complicates the test application. Testing delay faults in non-scan or partial scan design is further complicated by the fact that it is usually not practical to use a single fault assumption for delay faults. Therefore, in slow-fast-slow clock testing scheme it could happen that at the end of the fault activation phase more than one flip-flop latches a faulty value. The test generator has to account for this possibility in the fault propagation phase. Chakraborty et al. [24] consider different initial conditions for the fault propagation phase.

2.2.4 At-speed testing

At-speed testing strategy assumes that the fault is initialized, activated and propagated under a fast clock. Therefore, delay faults are present in all three phases.

At-speed testing strategies for sequential circuits have been proposed in [15, 16, 30, 119]. Pomeranz and Reddy [119] assume that multiple delay faults can simultaneously be present in the circuit and develop a value system for testing delay faults under these conditions. In their experiments several fast clocks (up to 3) were embedded in sequences of slow clocks. The at-speed test methodology proposed by Cheng [30] uses a single transition fault assumption (to be described in Chapter 3).

Faults that are untestable under the slow-fast-slow clock testing scheme remain untestable under the at-speed scheme. However, the converse may not be true [103].

2.3 TESTING HIGH PERFORMANCE CIRCUITS USING SLOWER TESTERS

Testing a design at its intended operating speed requires high speed testers. High cost of fast testers makes it impossible for the testers to follow the designs in terms of speed increase. The problem of testing high performance circuits without high speed testers has been addressed by a number of researchers [2, 6, 8, 10, 51, 58, 80, 149]. The proposed

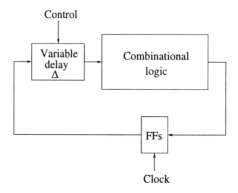

Figure 2.5. Inserting a controllable delay in the combinational logic.

strategies include tester pin multiplexing [2], built-in self-test [8], use of a high speed clock and shift registers [80], use of special test fixtures [10], reducing the supply voltage [58, 149], and use of on-chip test circuitry for testing high bandwidth memories [51].

The technique proposed by Agrawal and Chakraborty [6] involves adding extra logic to the combinational logic such that the speed of the circuit in the test mode becomes slower and comparable to the speed of the tester. The amount of the added delay can be controlled by a test input signal. Figure 2.5 illustrates the concept. The extra logic for inserting the variable delay should: (1) be controllable, (2) have a minimum normal mode delay, (3) be testable and (4) use minimum logic. The following example describes one possible implementation of such logic.

EXAMPLE 2.1 Consider the circuit in Figure 2.6(a) [6]. When the control input is set to 1, the input signal propagates to the output. When the control input is 0, the output holds its value. During normal operation, the control input is held at 1. If single clock master-slave flip-flops are assumed and the clock waveform is as shown in Figure 2.6(b), the falling edge A is the time when the

TEST APPLICATION SCHEMES FOR TESTING DELAY DEFECTS   15

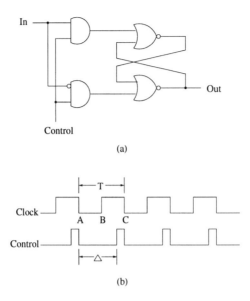

Figure 2.6. A controlled delay element and a waveform applied to the control input.

data is transferred from the master to slave flip-flop and the data stored in the slave flip-flop is applied to the combinational logic. The new data stored in the slave flip-flop will stay there until the next falling edge C. The rising edge B opens the master flip-flop to the input data. The time between the two falling edges (A and C) represents the clock period T. Figure 2.6(b) also shows the waveform for the control signal for the inserted logic. At the falling edge A of the clock, control signal drops to 0 and blocks the application of the data from the slave flip-flop to the combinational logic. After a delay Δ, the control signal rises to 1 and thus, allows the value from the slave flip-flop to be applied to the combinational logic. From Figure 2.6(b) it is clear that if T represents the clock period of the tester, then the clock period of the circuit can at most be $T_{rated} = T - \Delta$. In the test mode, delay Δ can be varied by changing the pulse width of the waveform.

The use of slow-fast-slow and at-speed testing schemes for testing high performance designs on slow testers has been discused by Krstić et al. [88]. The assumption is that the speed of the circuit is k (a positive integer) times higher than the speed of the tester and that an internal fast clock matching the speed of the circuit is available. If there is no fast clock available on the tester, the fast clock can be generated using frequency multiplier and the tester's clock.

2.3.1 Slow-fast-slow testing strategy on slow testers

The slow-fast-slow testing scheme can, under certain constraints, be used to test high performance circuits on low speed testers. In this scheme, the testable set of faults is affected by the presence or absence of latches on primary outputs. This is because to observe a fault, after activation, it has to be propagated to some primary output.

DEFINITION 2.1 Faults that in the activation time-frame can be propagated only to a primary output are called **PO-logic faults**.

DEFINITION 2.2 Faults that in the activation time-frame can be propagated to either a primary output or to a next state line and faults that in the activation time-frame can be propagated only to a next state line are called **NS-logic faults**.

Next, the use of slow-fast-slow scheme on slow testers for testing non-scan, scan and partial scan designs are considered.

Testing non-scan designs. The test application scheme for non-scan designs with latched PI/PO is shown in Figure 2.7(a). The primary inputs can be latched but it is not essential. The primary inputs are applied and the primary outputs are observed at the tester's speed. The tester's clock is also used in the slow phases (fault initialization and fault propagation). The tester's clock is assumed to be slow enough for the circuit to be fault-free in these phases. Fault activation is performed with a fast clock.

EXAMPLE 2.2 Consider the waveform in Figure 2.7(b). It illustrates the case when the tester's clock is 2 times slower than the operating speed of the circuit under test, i.e., $k = 2$. Also, it is assumed that the test sequence for the target fault consists of two initialization vectors (v_1 and v_2), one activation vector (v_3) and two propagation vectors (v_4 and v_5). Initialization vectors, v_1 and v_2 are applied at times t_1 and t_2, respectively. After the application of the activation vector at time t_3, the values of the primary outputs and next states are latched at time t_4. Next, the propagation vectors v_4 and v_5 are applied at times t_5 and t_6, respectively. Finally, at time t_7, the primary outputs are observed.

TEST APPLICATION SCHEMES FOR TESTING DELAY DEFECTS

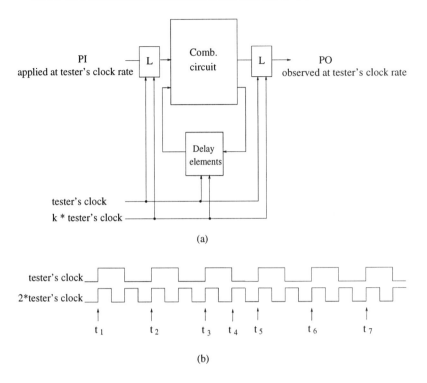

Figure 2.7. Non-scan designs with latched PI/PO.

Since the primary outputs can be latched at the end of the activation phase, this methodology can test both NS-logic and PO-logic faults.

When the primary outputs are not latched, PO-logic faults might not be testable on a slow tester using slow-fast-slow testing scheme. Only faults that are larger than a certain size can be tested. For example, PO-logic faults in the circuit in Figure 2.7(a) have to be larger than $t_5 - t_4$ to be testable.

Testing Scan Designs. Test application scheme that allows testing high speed scan designs on a low speed tester is illustrated in Figure 2.8(a). The tester's clock is used for applying the primary inputs, for the scan-in operation as well as for observation of the primary outputs and next state values, i.e., the scan-out operation. The fast clock is used for latching the values into primary outputs and next state lines.

EXAMPLE 2.3 The waveform in Figure 2.8(b) illustrates the case when $k = 2$. First the present state values, v_1, are scanned into the registers. Next, vector $v_1 = (i_1 + s_1)$ is applied at time t_1. If standard scan is used, the state values s_2 of the second vector can be obtained through functional justification [33]. The second vector, $v_2 = (i_2 + s_2)$, is applied at time t_2. Time $t_2 - t_1$ is assumed to be sufficient for the signal values to settle to their final values after the application of vector v_1 and before application of v_2. Next, one fast clock cycle is applied and at time t_3, the values of the primary outputs and next states are latched. At time t_4, the primary outputs can be observed and the scan-out operation can start. Then, the same cycle repeats for the next test.

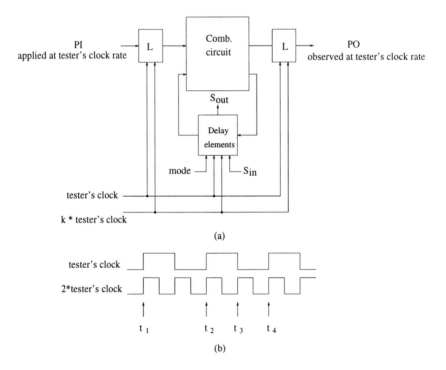

Figure 2.8. Scan designs with latched PI/PO.

Since the primary outputs can be latched after the application of the fast clock, both PO-logic and NS-logic faults can be tested using this scheme. However, if the scan circuit in Figure 2.8(a) does not have latches at the primary outputs, examining the waveform in Figure 2.8(b) we find that PO-logic faults must be larger than $t_4 - t_3$ to be testable.

Testing Partial Scan Designs. Testing scheme for partial scan designs represents a combination of the schemes described for non-scan and scan designs. The testing strategy depends on the target fault. For

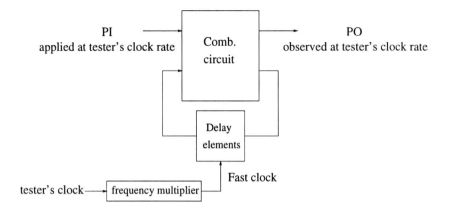

Figure 2.9. At-speed testing strategy for slow testers.

faults that can be tested through paths between the non-scan flip-flops, faults between non-scan flip-flops and POs and faults between PIs and non-scan flip-flops, the testing process is similar to the process described for faults in non-scan designs. It consists of initialization, activation and propagation phases. However, since some of the memory elements are scanned, the initialization and propagation phases might be shorter than in the non-scan case. For faults that can be tested through paths between the scanned flip-flops, faults between scanned flip-flops and POs and faults between PIs and scanned flip-flops, the testing strategy is the same as that described for scan designs.

2.3.2 At-speed testing strategy on slow testers

Conventional at-speed testing strategies for sequential circuits [30, 119] assume that the inputs are applied and the outputs are observed at the circuit's rated speed. This is impossible to do on a low speed tester. Krstić et al. [88] propose an at-speed method that accommodates the slow speed of the tester. Figure 2.9 illustrates the proposed at-speed scheme. The inputs to the circuit are applied and the outputs are observed at the slow tester's rate. Using the internal fast clock makes the circuit go through k states between applying the inputs and observing the outputs. This is equivalent to saying that the same set of primary input values are applied for k clock cycles and that the primary outputs are observed only at the end of each k-th cycle. Since the circuit runs at-speed between each application of inputs and observation of outputs, delay faults are constantly present in the circuit.

20   CHAPTER 2

EXAMPLE 2.4 Figure 2.10(a) illustrates the proposed at-speed testing scheme for $k = 3$. The same set of primary input values is applied for three fast clock cycles and the primary outputs are only observed after the third cycle. The delay elements are clocked with the fast clock and the circuit passes through three different states before the application of the next primary input vector.

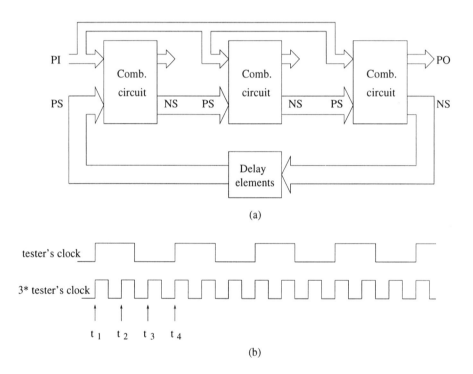

Figure 2.10.   At-speed testing scheme for $k = 3$.

Since the observation of the outputs is performed at the tester's speed, the existence or non-existence of latches at the primary outputs does not affect which faults can be tested using this at-speed scheme. This means that the proposed at-speed scheme can be used to test PO-logic faults that cannot be tested using slow-fast-slow scheme. Input and output clocks should be skewed to test PI to PO paths terminating at the unlatched POs.

Next, the application of the at-speed testing scheme to non-scan, scan and partial scan designs is considered.

Testing Non-Scan Designs. Let the design in Figure 2.10(a) be a non-scan design and consider the waveform shown in Figure 2.10(b). At time t_1, vector $v_1 = (i_1 + s_1)$ is applied to the circuit. Next, at time t_2, the primary input values stay unchanged but the state values have changed. Therefore, at time t_2, vector $v_2 = (i_1 + s_2)$ is applied to the circuit. Similarly, at time t_3, vector $v_3 = (i_1 + s_3)$ is applied. Finally, at time t_4, the primary outputs can be observed. A new vector, $v_4 = (i_2 + s_4)$, is applied to the circuit at time t_4 and the cycle repeats. In this scheme, if the test sequence contains n test vectors, where n is a positive integer, the circuit actually changes $k \times n$ states. For example, the circuit in Figure 2.10(a) must go through 3 or 6 or 9, ... states. Therefore, the test generation process for this at-speed testing strategy has to be different than the test generation process for at-speed schemes that assume fast testers.

Testing Scan Designs. Let the design in Figure 2.10(a) be a scan design and consider the waveform shown in Figure 2.10(b). The application of primary inputs, scan-in, scan-out and observation of the primary output values are performed at the tester's speed. However, between the scan-in and scan-out operations, the circuit is allowed to run with the fast clock and it goes through three states while the primary inputs are kept constant. At time t_1, the first set of state values, s_1, is assumed to be already scanned-in and i_1 is applied at the primary inputs. The state values for the second and third vector, $v_2 = (i_1 + s_2)$ and $v_3 = (i_1 + s_3)$, are obtained through functional justification and these vectors are applied at times t_2 and t_3, respectively. At time t_4, the values of the primary outputs are observed and the scan-out operation starts. The test sequence for scan designs contains k vectors.

Testing Partial Scan Designs. As with slow-fast-slow scheme, the at-speed testing strategy for partial scan designs can be described as a combination of testing strategies for scan and non-scan designs (depending on the target fault).

Since in this at-speed testing strategy the primary outputs were observed only after each k-th cycle, the signal observability is smaller than that obtained if the primary outputs are observed after each cycle. Also, since the primary inputs are kept unchanged for k clock cycles, the controllability of the signals is negatively affected as well. This can lead to a lower fault coverage than that obtained with a high speed tester. Therefore, the described at-speed technique should not be used as a stand-alone technique. Instead, it should be combined with the slow-fast-slow testing strategy to obtain higher overall fault coverage. When there are no latches on primary outputs, the at-speed technique can be used to test PO-logic faults that would stay untestable under the slow-fast-slow strategy. In addition, some NS-logic faults might also be untestable by the slow-fast-slow scheme but testable by the at-speed scheme. The proposed at-speed scheme can be used to detect them as well. Also, there exist faults that can be tested by both slow-fast-slow testing strategy and by the at-speed strategy. If these two strategies require that the circuit passes through a comparable number of states when testing a given fault, the at-speed scheme would clearly be superior in terms of the testing time.

2.4 SUMMARY

Test application strategy is an integral part of delay test generation. This is especially true for testing sequential designs for which several different strategies exist. Enhanced and standard scan schemes allow trade-offs between overheads (area and test application time) and fault coverage. Enhanced scan requires high area and test application time overheads but results in a higher fault coverage than standard scan techniques. In slow-fast-slow testing scheme, the assumption that the circuit is fault-free in the fault initialization and propagation phases greatly reduces the complexity of test generation but it complicates the test application process when compared to the at-speed testing scheme.

An important factor in delay fault test generation is also the tester's speed. The speed of the testers usually lags behind the speed of the new designs. Therefore, developing new techniques that would allow testing high speed designs on slower testers is of great practical importance.

3 DELAY FAULT MODELS

The focus of this chapter is on the ways to model delay faults. Five delay fault models are considered: transition fault model, gate delay fault model, line delay fault model, path delay fault model, and segment delay fault model. It is assumed that in the nominal design each gate has a given fall (rise) delay from each input to the output pin. Also, the interconnects are assumed to have given rise (fall) delays. Since the gate pin-to-pin delays and the interconnect delays can be combined together, the term "gate delay" will be used to denote this sum. Transition, gate and line delay models are used for representing delay defects lumped at gates while the path and segment delay models address defects that are distributed over several gates. The advantages and disadvantages of each model are discussed.

3.1 TRANSITION FAULT MODEL

Transition fault model [30, 94, 135, 151] assumes that the delay fault affects only one gate in the circuit. There are two transition faults associated with each gate: a slow-to-rise fault and a slow-to-fall fault. It is assumed that in the fault-free circuit each gate has some nominal delay.

Delay faults result in an increase or decrease of this delay. (Throughout this book only delay faults caused by an increase of the delay will be considered.) Under the transition fault model, the extra delay caused by the fault is assumed to be large enough to prevent the transition from reaching any primary output at the time of observation. In other words, the delay fault can be observed independent of whether the transition propagates through a long or a short path to any primary output. Therefore, this model is also called the *gross delay fault model* [112]. In addition to being a model for delay faults, transition fault model is also used as a logic model for transistor stuck-open faults in CMOS circuits [148]. The CMOS transistor stuck-open faults can be treated as faults that either suppress or delay the occurrence of certain transitions. In practice, the extra delay caused by a stuck-open fault depends on the electrical characteristics of the defective component.

To detect a transition fault in a combinational circuit it is necessary to apply two input vectors, $V = \langle v_1, v_2 \rangle$. The first vector, v_1, initializes the circuit, while the second vector, v_2, activates the fault and propagates its effect to some primary output. Vector v_2 can be found using stuck-at fault test generation tools. For example, for testing a slow-to-rise transition, the first vector initializes the fault site to 0, and the second vector is a test for stuck-at-0 fault at the fault site. A transition fault is considered detected if a transition occurs at the fault site and a sensitized path extends from the fault site to some primary output.

The fault equivalence rules for transition faults are more restrictive than those for stuck-at faults [151]. This is because, as mentioned above, testing a transition fault requires more than one vector. Only two rules can be applied for fault equivalence collapsing for transition faults: (1) if a gate has a single input, then the input transition faults are equivalent to the output transition faults, and (2) if a gate has only one fanout, then the output transition faults are equivalent to the input transition faults on the fanout gate. As a result, the number of collapsed transition faults for a given circuit is larger than the number of collapsed stuck-at faults.

The main advantage of the transition fault model is that the number of faults in the circuit is linear in terms of the number of gates. Also, the stuck-at fault test generation and fault simulation tools can be easily modified for handling transition faults. On the other hand, the expectation that the delay fault is large enough for the effect to propagate through any path passing through the fault site might not be realistic because short paths may have a large slack. The assumption that the delay fault only affects one gate in the circuit might not be realistic, either. A delay defect can affect more than one gate and even though none of the individual delay faults is large enough to affect the performance of the circuit, several faults can together result in a performance degradation. For practical simplicity, the transition fault model is frequently used as a qualitative delay model and circuit delays are not considered in deriving tests.

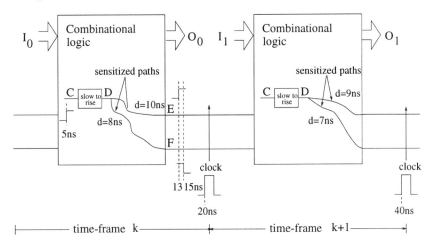

Figure 3.1.   Faults of different size result in different next states.

Transition fault model for sequential circuits. The transition fault model described above cannot be used for sequential circuits if the clock is applied at the rated speed because it does not take into account the fault size. Next, we discuss a transition fault model [30] that is suitable for the at-speed test application scheme.

The transition fault model for a sequential circuit [30] is characterized by the fault site, the fault type and the fault size. As before, the fault type is slow-to-rise or slow-to-fall transition. The fault size represents the amount of extra delay caused by the defect. In sequential circuits, different fault sizes will result in different faulty next states.

EXAMPLE 3.1 Consider the circuit in Figure 3.1. It shows two time-frames of a sequential circuit. It is assumed that input vectors are applied at the rated speed. The clock pulse for latching the next state is applied before the next input vector is applied. Suppose there is a slow-to-rise fault between the signal C and signal D. The clock interval is 20 nanoseconds (ns). The inputs are applied at 0ns (reference time) for time-frame k and a rising transition occurs at signal C at 5ns. There are two sensitized paths from C to the next state signals, E and F. The propagation delays of the transitions along these two paths are 10 and 8 ns, respectively. The transitions at E and F for the fault-free circuit and the times at which they occur are shown in the figure. If the fault size of the slow-to-rise fault at C is less than 5ns, the next state of the faulty circuit will be the same as that of the fault-free circuit, i.e., $(E, F) = (1, 0)$. If the fault size is greater than 5ns but less than 7ns, flip-flop E will catch the fault effect but flip-flop F will not (when the clock is applied at 20ns). The faulty next state will be $(E, F) = (0, 0)$. If the fault size is greater than 7ns, the faulty next state will be $(E, F) = (0, 1)$. The faulty next state, along with the next input vector, will produce a new logic value at each signal in time frame $k+1$. Next, let the longest and shortest sensitized paths from C to any next state signal in time-frame $k+1$ have delays of 9ns and 7ns, respectively. If the value at C in time-frame $k+1$ is a logic 1 and if the fault size is in the range of (7ns, 26ns), the value at signal D in time frame $k+1$ will be 1. The effects of the delayed transition will be stabilized at the next state signals in time-frame $k+1$ before the following clock pulse is applied at 40ns. On the other hand, if the new value at C is a logic 0, regardless of the fault size, the delayed transition will be completely suppressed and the value at D in time-frame $k+1$ will be 0. If the value at C in time-frame $k+1$ is a logic 1 and if the fault size is in the range of (26ns, 28ns), the effects of the delayed transition will be propagated to signal E and stabilized before the next clock pulse is applied. However, the next state signal F will not catch the fault effect in this case.

Clearly, it is not possible to guarantee the detection of a transition fault in a sequential circuit under the at-speed test application scheme without considering the size of the fault. Different fault sizes result in completely different circuit behaviors. However, the computation costs of dividing the fault sizes into hundreds of fine-grained ranges and simulating them are prohibitive. This problem can be solved by dividing the fault size using units of clock cycles [30]. A fault can then be specified as a triple (s, t, i) where s is the faulty signal, t is either slow-to-rise or slow-to-fall and i is an integer specifying the fault size in units of clock cycles.

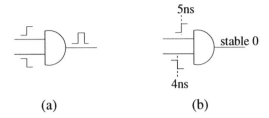

Figure 3.2. Advantage of considering circuit timing.

3.2 GATE DELAY FAULT MODEL

Gate delay fault model [22, 69, 70, 71, 126, 128] assumes that the delay fault is lumped at one gate in the circuit. However, unlike the transition model, the gate delay fault model does not assume that the increased delay will affect the performance independent of the propagation path through the fault site. It is assumed that long paths through the fault site might cause performance degradation. The gate delay fault model is a quantitative model since it takes into account the circuit delays. The delays of the gates are represented as either the worst-case values or intervals.

Taking the timing into consideration when deriving tests for gate delay faults allows application of some tests that would otherwise not be considered.

EXAMPLE 3.2 Consider the AND gate shown in Figure 3.2. If no information about the delays in the circuit is available, it might be assumed that there is a static hazard at the output of the AND gate, as shown in Figure 3.2(a). This static hazard might prevent the propagation of some target fault elsewhere in the circuit. However, if the information about delays is given and the arrival times of the transitions at the inputs to the AND gate are as shown in Figure 3.2(b), the output will clearly have a stable 0 value, which may be favorable for the propagation of the target fault effect.

To determine the ability of a test to detect a gate delay defect it is necessary to specify the delay size of the fault. Methods for computing the smallest delay fault size (detection threshold) guaranteed to be detected by some test have been reported in the literature [69, 70, 126, 128].

The limitations of the gate delay fault model are similar to those for the transition fault model. Because of the single gate delay fault assumption a test may fail to detect delay faults that are result of the sum of several small delay defects. The main advantage of this model is that the number of faults is linear in the number of gates in the circuit.

3.3 LINE DELAY FAULT MODEL

A variation of the gate delay fault model is the line delay fault model. **Line delay fault model** [101] tests a rising (falling) delay fault on a given signal (line) in the circuit. The fault is propagated through the longest *sensitizable* path passing through the given line. Similar to the transition and gate delay fault models, line delay fault model assumes a single delay fault. Therefore, the number of faults equals twice the number of lines in the circuit. Sensitizing the longest path through the target line allows detecting the delay fault of the smallest size on the target line. In general, a test will cover several line delay faults. Therefore, this fault model can detect some distributed delay defects on the propagation paths. However, since only one propagation path through each line is considered, it may fail to detect some defects [100].

3.4 PATH DELAY FAULT MODEL

Under the **path delay fault model** [139] a combinational circuit is considered faulty if the delay of any of its paths exceeds a specified limit. A path is defined as an ordered set of gates $\{g_0, g_1, \ldots g_n\}$, where g_0 and g_n are a primary input and primary output, respectively. Also, gate g_i is an input to gate g_{i+1} ($0 \leq i \leq n-1$). A delay defect on a path can be observed by propagating a transition through the path. Therefore, a path delay fault specification consists of a physical path and a transition that will be applied at the beginning of the path. The delay or length of the path represents the sum of the delays of the gates and interconnections on that path.

Tests for the path delay fault model can detect small distributed delay defects caused by statistical process variations. A major limitation of this fault model is that the number of paths in the circuit can be very large (possibly exponential in the number of gates). For this reason testing all path delay faults in the circuit is not practical. Two strategies are commonly used for selecting the set of path delay faults for testing. One is to select a minimal set of paths such that for each signal s in the circuit the longest path containing s is selected for testing [95, 101, 104]. The other is to select all paths with expected delays greater than a specified threshold. The reason behind selecting the longest paths is that the delay defects on shorter paths might not be large enough to affect the circuit performance. Also, if the defects on short paths are large and could affect the performance, one expects that such defects would be detected by other tests (e.g., at-speed tests and gate delay tests) that precede the path delay fault testing. This strategy might work for circuits whose paths have very different delays so that there is a small percentage of long paths. However, often in performance optimized designs almost all paths have long delays and in these circuits not even all longest paths can be tested [113]. Therefore, even after path delay fault testing, the temporal correctness of the circuit under test often cannot be guaranteed. The problem can be alleviated by developing techniques for resynthesizing the circuits such that the path count is reduced [85, 125]. These techniques will be described in Chapter 8.

Additional problems with the use of the path delay fault model are: (1) Tests that guarantee that the given path will not affect the performance of the circuit can be generated using reasonable resources only for a small subset of paths in the circuit. For most circuits, there exists a large number of paths that can impact the performance of the circuit but these paths cannot be easily tested. Classification of path delay faults based on their testability characteristics is considered in Chapter 5. (2) Most path delay fault testing research has concentrated on testing combinational circuits. Extending these techniques to non-scan or partial scan designs is not straightforward.

3.5 SEGMENT DELAY FAULT MODEL

Segment delay fault model [64, 65] represents a trade-off between the transition delay fault model and the path delay fault model. The assumption in this model is that the delay defect affects several gates in a local region of occurrence. Also, it is assumed that a segment delay fault is large enough to cause a delay fault on all paths that include the segment. The length of the segment, L, can be anywhere from 1 to L_{max}, where L_{max} represents the number of gates in the longest path in the circuit. The fault list consists of all segments of length L and all paths whose length is less than L. When $L = 1$, this model reduces to the transition fault model. When $L = L_{max}$, the segment delay fault model is equivalent to the path delay fault model. The idea of using the segment delay model is to combine the advantages of the transition and path delay fault models while avoiding their limitations. Since the number of segment delay faults for a given L can be much smaller than the number of all paths in the circuit, the explosion of the number of faults can be avoided. Also, the assumption that the fault is distributed over several segments is more realistic than the transition fault assumption about the lumped delay fault at one segment. In practice, many segments are testable while the entire paths containing those segments may not be testable.

The length of the segment can be decided on the basis of available statistics about manufacturing defects. All segments of a given length can be counted and identified using the method in [64].

3.6 SUMMARY

Fault models represent an approximation of the effects that defects produce in the behavior of the circuit. An ideal model should provide a high confidence level that faulty circuits will be isolated. The test generation process for such a fault model should allow handling of very large designs with reasonable amount of computing resources. Detecting timing

Table 3.1. Comparison of different delay fault models.

Delay fault model	number of faults w.r.t. number of gates	faults that can be tested	size of detectable faults	test generation
transition	linear	lumped at gate	large	modified stuck-at ATPG
gate	linear	lumped at gate	larger than threshold	takes timing into account
line	linear	lumped at gate or distributed along certain paths	small to large	requires finding longest sensitizable path through line
path	exponential (worst case)	distributed along paths	small to large	hard
segment	linear to exponential	distributed along segments	small to large	depends on the segment length

defects requires models other than the well known stuck-at fault model. Several different delay fault models have been proposed in the literature. Each of these models has its advantages and disadvantages. The main characteristics of delay fault models are summarized in Table 3.1. The path delay fault model is usually considered to be closest to the ideal model for delay defects. However, testing all path delay faults that can affect the performance of the circuit is impractical. Currently used path delay fault model is an oversimplification for deep submicron devices for which the interconnect and cell delays are highly pattern dependent. Developing a more accurate fault model and selection of critical paths in new designs that are highly sensitive to process variations, circuit defects and signal coupling effects are important research problems for the future.

4 CASE STUDIES ON DELAY TESTING

This chapter addresses the motivation behind subjecting a VLSI design to delay testing. The need for delay testing can best be demonstrated by experimental data on real designs. Therefore, this chapter will focus on several case studies of delay testing conducted by research groups in industry and academia [12, 13, 21, 107, 108, 109, 147].

Experience has shown that stuck-at fault testing is not sufficient to guarantee high product quality requirements. There exist some faults that can be detected only if multiple test strategies are used, e.g., stuck-at, I_{DDQ}, delay testing, etc. An experiment performed at IBM has demonstrated that "randomly occurring gross delay defects can allow chips to pass full stuck-at fault testing at both wafer and module levels, but cause them to fail when operated at system speeds" [21]. In that experiment, 60,000 modules from an IBM computer system were subjected to delay testing. The modules were designed as CMOS standard cell/gate array devices and the experiment was performed using the transition fault model. These modules had passed the slow speed tests for stuck-at faults. The transition fault coverage of the delay test was 75%. There were 97 modules that failed delay testing, i.e., these modules had gross delay defects not detected by the stuck-at fault testing. The timing characterization of the faulty modules showed that the defect size was between 0 and 1900 ns, where defect size represents the amount of propagation delay added by the fault. The clock cycle time was 120 ns. Figure 4.1 [21] gives the distribution of defect sizes for 91 out of 97 modules that failed the delay test. As the histogram shows,

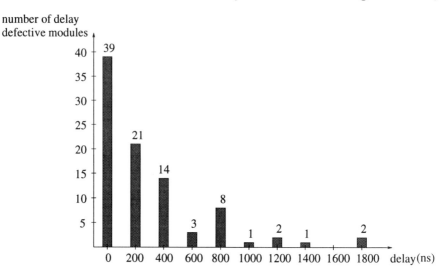

Figure 4.1.   Fault size distribution in an IBM experiment.

39 (or 42.8%) modules had delay defects of sizes between 1 and 100 ns, while 60 (or 66%) modules had delay defects of sizes between 1 and 200 ns. The failure analysis identified resistive first-layer-metal opens as the dominant cause of the delay defects. This experiment clearly shows the value and need for delay testing.

The studies by Maxwell et al. [107, 108] considered the detection of timing defects by functional and I_{DDQ} tests. These studies have shown that functional tests applied at-speed and I_{DDQ} tests can detect some, but not all, delay defects. Their experiment [108] analyzed a sample of three wafer lots consisting of 26,415 die (fully static standard cell design with 8,577 gates and 436 flip-flops). The test set consisted of functional vectors run at slow speed (2 MHz) and at high speeds (20 MHz and 32 MHz), scan tests and I_{DDQ} tests. These tests identified 4,349 devices as faulty. Figure 4.2 [108] illustrates the distribution of the failing die in each test class. As it can be seen, each test set detected some failures not detected by any other test. In this experiment, a total of 21 devices passed scan and low speed (2 MHz) functional tests but failed the at-speed (20 and 32 MHz) functional tests. Only 10 out of these 21 modules

CASE STUDIES ON DELAY TESTING   35

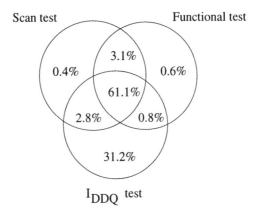

Figure 4.2.   Distribution of failing die.

with delay defects were isolated by I_{DDQ} tests. Therefore, to improve the product quality level, tests with high static stuck-at coverage should be combined with I_{DDQ} and delay tests [108].

The improvement of the product quality level obtained by including delay tests has been studied by Maxwell *et al.* [109]. The delay tests were generated based on the transition fault model but some path delay fault tests were also included. A large sample of modules from three wafer lots were tested. Experiment was performed on a fully static 25k-gates standard cell design with 1,497 flip-flops (full scan design), fabricated by a 0.8 μm process with 3 metal layers. The design uses a single 33MHz clock. The test set combined functional, stuck-at fault, I_{DDQ}, transition fault and path delay fault vectors. Figure 4.3 [109] shows the distribution of failing modules when functional, stuck-at and delay tests were applied. As can be seen, delay tests rejected 24 modules that passed other tests. Figure 4.4 [109] shows the distribution of failing modules by test sets consisting of at-speed functional, stuck-at scan, I_{DDQ} and delay test vectors. The results are given for two different detection thresholds for I_{DDQ} tests: 50 μA and 200 μA. Again, the delay test isolated a number of faulty modules that passed all other test sets. As the diagrams show, the number of faulty modules isolated only by delay tests increases when the I_{DDQ} threshold is increased from 50 μA to 200 μA. This indicates that, in case of deep sub-micron devices for which increased leakage current might require raising the I_{DDQ} threshold, application of delay tests will become even more important [109]. In this experiment, tests for path delay faults have isolated only 18.5% of the devices with delay defects. This poor performance of the path delay

36   CHAPTER 4

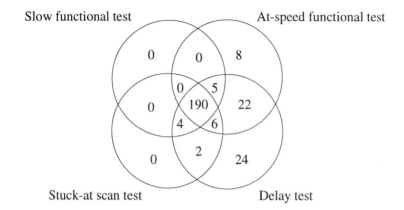

Figure 4.3.   Distribution of failing modules without applying I_{DDQ} tests.

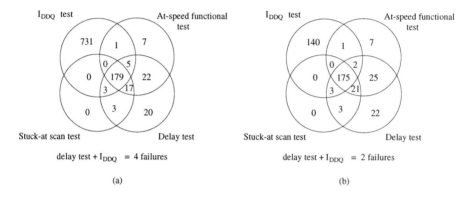

Figure 4.4.   Distribution of failing modules with I_{DDQ} threshold set to (a) 50 μA and (b) 200 μA.

fault tests can be explained on the basis of the selection of paths for testing in this experiment. Only a subset of longest paths was selected for testing. Static timing analysis was used to extract long paths from the netlist. The analysis of 2,000 longest paths showed that a majority of them could not be sensitized such that the detection of the fault on them would be guaranteed independent of the signal delays outside the target path. These paths might have been the cause of timing failures and generating tests under different sensitizing conditions might have improved the fault coverage [34]. Out of the 2,000 long paths only 148 were selected for testing. Due to tool limitations only paths between flip-flops could be considered in this experiment. This might have also contributed to the low path delay fault coverage because the analysis has shown that most of the failures identified by the transition fault tests involve pads.

Testing resistive input bridging, short and open faults by delay and I_{DDQ} tests has been investigated by Vierhaus et al. [147]. The study included detailed simulations of resistive stuck-on, stuck-open and bridging faults in typical static CMOS circuits. For example, Figure 4.5 illustrates the simulated faults in a 2-input CMOS AND gate. In this

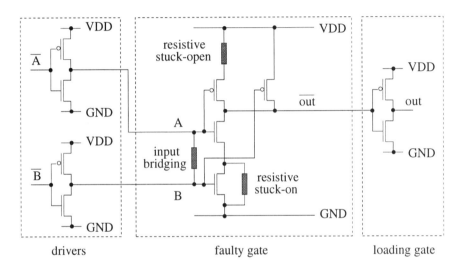

Figure 4.5. Considered resistive faults in a 2-input CMOS AND gate.

study, it was assumed that the resistive faults only exist in the 2-input NAND circuit. However, the characteristics of the driving and loading stages have an impact on the fault behavior. Therefore, delay and I_{DDQ} effects were simulated for the 3-stage units, taking into account the delays and overcurrents of the driving and loading stages. The summary of the delay and I_{DDQ} effects for the 2-input AND gate in 1.5 – 2 micron CMOS technology is shown in Figure 4.6. Both resistive input bridges and resistive transistor stuck-on faults resulted in a stuck-at fault for resistance values of up to 5 kΩ. There were no gross delay effects (delay larger than 10 gate delays) for these faults. Therefore, resistive input bridges and resistive transistor stuck-on faults cannot be safely detected using tests for transition faults. Using tests for path delay faults might have helped since they can detect small delay defects like those caused by resistor values in the 10 to 100 kΩ range. Both resistive input bridges and resistive transistor stuck-on faults have resulted in a gross overcurrent (larger than 100 μA) for the resistance values of up to 100 kΩ.

Therefore, resistive input bridges and resistive transistor stuck-on faults can be easily detected using I_{DDQ} tests. Resistive transistor stuck-open faults in the 100 kΩ to several MΩ range cause gross delay effects. These faults cannot be detected by I_{DDQ} tests.

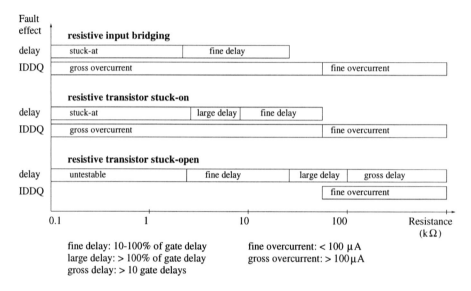

Figure 4.6. Delay and I_{DDQ} effects for resistive faults in a 2-input CMOS AND gate.

Evaluating multiple testing strategies through experiments was also pursued by Franco et al. [47, 48] and Ma et al. [99]. In those experiments a CMOS gate array test chip with 25k-gates was designed, manufactured and subjected to different types of test vectors. Each chip included 4 copies of 5 different types of circuits and test support circuitry. The circuits were limited to 24 inputs in order to allow the application of exhaustive tests that would serve as a reference for other tests. The 5 circuit types included two multipliers and three control blocks. The three control blocks represented three different implementations of the same function. The applied test sets included tests for single stuck-at faults, transition faults, gate delay faults, path delay faults as well as 2^n exhaustive test (test consisting of all possible combinations of one-pattern tests for n inputs). Before application of these tests, the design was subjected to gross parametric and test support circuitry tests and the failing dies were removed. The experiment investigated "at-speed" and "delay" clocking modes. In the first case all vectors were applied at the rated speed while in the second case, the signals were allowed to settle down after the application of the first vector and before the application of the second vector (see Chapter 2, Figure 2.1). Also, the application of the test sets was done at three different speeds: rated speed, slower than the rated speed (2/3 of the rated speed) and faster than the rated speeds (25% faster for the multiplier circuits and 5% faster for the control logic). A total of 128 circuits failed at least one of the applied test sets at rated or slow speed, while 1 circuit failed at the fast speed. The fact that only one circuit failed at the speed faster than the rated speed indicates that the detected timing failures are true delay defects and not just a result of aggressive timing [48]. Since each test was repeated under several different conditions, the collected information about failures allowed classification of defects into categories.

DEFINITION 4.1 [48] A **timing-independent combinational (TIC) defect** is a defect whose behavior is independent of (1) the input patterns applied prior to the current input pattern, and (2) the clock speed (rated speed or slower than the rated speed).

The **non-TIC defects** depend either on the clock-speed or on the previously applied patterns or on both. Figure 4.7 shows the defect classification. Out of 128 defective circuits, 72 were classified as having TIC

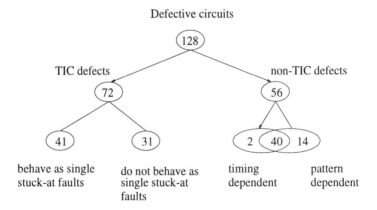

Figure 4.7. Defect classification.

defects while 56 had non-TIC defects. There were 41 circuits with TIC defects that behaved as single stuck-at fault defects. Out of 56 circuits with non-TIC defects, 42 were found to depend on the clock rate (timing dependent) and 54 were found to be pattern dependent. Therefore, timing or pattern dependent defects represented 44% of all the defects in this experiment. Table 4.1 shows the test results for the circuits that have escaped at least one test. The results are given for two transition

Table 4.1. Distribution of escapes for different test sets.

| Test set | Num. of defective circuits tested | Number of escapes |||||
| | | at-speed || delay |||
		s	r	s	r	f
Transition fault (Test 1)	128	10	6	6	4	5
Transition fault (Test 2)	128	12	11	10	8	7
Gate delay (X → 0)	88	17	16	18	16	16
Gate delay (X → rand)	88	13	11	11	9	9
Path delay (Crit. path, X → 0)	128	35	35	35	34	34
Path delay (Crit. path, X → rand)	128	27	27	28	27	27
Path delay (Robust, X → 0)	69	7	5	4	2	1
Path delay (Robust, X → rand)	69	8	5	5	3	2
Path delay (Robust)	69	7	4	3	2	2
Path delay (Non-Robust, Test 1)	69	12	10	10	7	7
Path delay (Non-Robust, Test 2)	69	8	4	4	2	2
Single stuck-at (Test 1, 100%)	128	14	10	10	9	8
Single stuck-at (Test 2, 100%)	128	9	7	7	5	3
Pseudo-random/Exhaustive	128	7	2	3	1	1

X → 0: X is replaced by 0
X → rand: X is replaced randomly by 0 or 1

fault tests (generated by two different tools), two gate delay fault tests (one in which the unknown values were replaced by 0's and the other in which the unknown values were replaced randomly by 1's or 0's), two path delay tests containing only tests for the longest paths, three robust path delay fault tests, two non-robust path delay fault tests, two stuck-at fault tests with 100% fault coverage and a pseudo-random test which is also an exhaustive test. The second column shows the number of circuits to which each test set was applied. Some of the test sets were applied only to a subset of defective circuits. Columns 3 and 4 show the results for the "at-speed" clocking mode for the case of slow (s) and rated (r) clock speeds, respectively. The last three columns show the results for "delay" clocking mode for slow, rated and fast clock speeds, respectively. Path delay fault tests targeting only the longest paths had the worst performance among all test sets. The critical path delays were calculated using the pre-layout delay values and the delay inaccuracies might have contributed to the poor performance of these tests. For all test sets, testing at slower speed resulted in more escapes. No circuit passed the rated speed and failed the slow speed testing [99]. For the same clock speed, the "delay" clocking mode had fewer escapes than the "at-speed" clocking mode. Franco et al. [48] explain this by the fact that the timing defects caused by extra capacitive coupling or ground bounce might be better activated by a test applied at-speed, while the defects that depend on charging or discharging of capacitances might be better activated by a test that allows complete discharging of the capacitor before it starts charging again. The experiment has also shown that for single stuck-at tests and for the exhaustive test most of the test escapes are for the circuits with non-TIC defects. This, again, indicates that detecting timing failures requires deterministic test generation.

The experiment by Franco et al. [48] has also investigated the issue of delay modeling. The propagation delay was measured on four die for a 12-input multiplier during the application of different test sets. Figure 4.8 shows the test setup for measuring propagation delays [48]. The

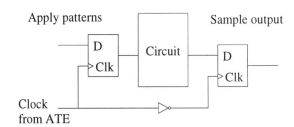

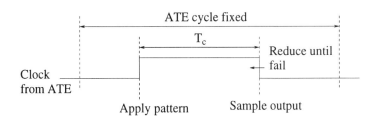

Figure 4.8. Test setup for measuring the propagation delays.

patterns were applied to the circuit on the rising edge of the clock and the outputs were observed on the falling edge. To measure the propagation delay, the clock cycle was reduced until the circuit failed the test. The experiment was repeated for several test sets, including tests for 100 critical path delay faults. Testing all path delay faults was impossible since there were 7×10^{15} paths in the circuit. The critical paths were chosen using a unit delay model. A 2^{2n} super-exhaustive test (test that contains all possible combinations of two-pattern tests for n inputs) was also applied. It was found that the longest propagation delay was exercised only by the super-exhaustive test. Table 4.2 shows the comparison of the maximum and minimum values of the propagation delay for the four tested die relative to those obtained for super-exhaustive tests. For

Table 4.2. Comparison of delay propagation measurement results.

Test set	Number of patterns	Number of strobes	Measured propagation delay relative to longest delay	
			min	max
Single stuck-at fault	22	22	0.88	0.90
Transition fault	304	152	0.85	0.93
Critical path, $X \to 0$	1692	846	0.88	0.91
Critical path, $X \to$ rand	1692	846	0.87	0.91
Critical path, $X \to$ rand	1692	1692	0.90	0.91
Gate delay, $X \to 0$	976	488	0.78	0.85
Gate delay, $X \to$ rand	976	516	0.78	0.85
Super-exhaustive	16.8M	16.8M	1.00	1.00

$X \to 0$ – X is replaced by 0
$X \to$ rand – X is replaced randomly by 0 or 1

some tests the outputs were sampled after each vector while for others, the outputs were sampled after every other vector (as shown in columns 2 and 3). The results for the propagation delays are similar for single stuck-at fault tests and critical path tests (the circuit should be tested at a rate that is about 10% higher than the rated speed in order to detect timing failures). This illustrates the danger of selecting only a small set of paths and using inaccurate delay values for critical path selection [48].

Another study on the effectiveness of delay testing was conducted at Bell Labs by Bose et al. [12, 13]. They simulated the ISCAS'89 benchmark s1196 for two sets of vectors using a timing simulator [5]. Rise and fall delays of gates were modeled for a CMOS standard cell library. Vectors were generated in the non-scan sequential mode. One set of vectors was generated specifically to cover stuck-at faults [11]. The other set was designed to cover path delay faults in non-robust mode [4].

Path delay fault simulation [14] showed that the longest path covered by delay test vectors had 24 gates. The longest path in the circuit, however, had 25 gates. Such simulation is independent of delays in the circuit. Both robust and non-robust modes tested some paths that had 24 gates although comparatively fewer paths were covered in the robust mode. The longest path tested by the stuck-at fault vectors had 19 gates for the non-robust mode and 16 gates for the robust mode.

Bose et al. further performed multiple-delay timing simulation of both sets of vectors. By gradually reducing the vector application period, they determined the largest clock period for which the tests showed failures at primary outputs. The expected (correct) response was the simulated output for a vector application period larger than the longest path delay. They found that the clock frequency at which vectors are applied has a strong influence on the delay test effectiveness.

DEFINITION 4.2 **Guaranteed failure frequency (GFF)** for a given set of vectors is the lowest clock frequency above which the circuit will definitely show a failure if any delay fault detectable by the vectors exists [13].

Notice that GFF is a characteristic of the vector set which also depends on the nominal delay of the longest sensitized and tested path. The nominal delays are usually estimated or measured and are used as design parameters. Bose et al. determined the GFF by selectively suppressing hazards in their simulator. Without any hazard suppression, both delay and stuck-at vectors showed failures at all clock periods smaller than 105 ns. With hazard suppression the GFF for delay test vectors was found to be 9.52MHz, which still corresponds to a clock period of 105 ns. However, for the stuck-at fault vectors the GFF was 14.29MHz (70 ns clock period).

Even though the stuck-at fault vectors produced a faulty output at 9.52MHz clock, that fault was due to the propagation of a static hazard. The delay test vectors, on the other hand, produced the failure by propagating a transition. The longest path through which the stuck-at fault vectors propagated a transition had a delay of only 70 ns. Bose *et al.* suggest that the vectors should be applied using the clock rate of the guaranteed failure frequency. This, of course, is a suitable procedure if we assume delay variations to be correlated over the entire circuit. In general, for given choices of vector sets the one with lowest GFF should be used for delay testing. Obviously, the tests derived for stuck-at faults did not fare well in this experiment.

4.1 SUMMARY

As experimental results show, describing a defect behavior using only the stuck-at fault model can result in many undetected defects. A large portion of the undetected defects represents timing failures. Some of these timing failures can be detected by functional vectors applied at-speed or by I_{DDQ} vectors. Transition fault tests have been shown to be effective for detecting gross delay defects that would otherwise remain undetected. For high performance circuits with aggressive timing requirements, small process variations can lead to failures at the system clock rate. These defects can be detected using tests for path delay faults. While the experiments clearly demonstrate the need for detecting delay defects, the high cost for detecting them is still a serious problem. A possible cost effective strategy for delay testing would include:

- use of functional vectors that could be applied at-speed and should catch some delay defects. Functional vectors should be evaluated for transition fault coverage.
- application of deterministic tests for undetected transition faults
- application of deterministic tests for long path delay faults.

Selection of critical paths has been shown to have a significant effect on the detection of timing failures. Moving towards deep submicron designs puts even more emphasis on a proper selection of critical paths, based on accurate gate and interconnect delay information and noise factors such as crosstalk. Also, studies have shown that in most designs critical paths cannot be sensitized using propagation conditions that would guarantee the detection of a fault independent of the delays of signals outside the target path. Including tests for such paths can further improve the product quality level.

5 PATH DELAY FAULT CLASSIFICATION

This chapter is devoted to a discussion on the classification of the path delay faults. Paths are classified according to their testability characteristics. A given path delay fault can be tested by many different tests. Unlike a stuck-at fault for which all tests have the same quality (fault is certainly detected by the test), in path delay fault testing different tests for a given fault have different levels of quality (probability of detection). For example, some tests can guarantee detection of a fault while others can detect the fault only under restricted conditions. Not every path can be tested with a highest quality test. This is because higher quality path delay fault tests require more stringent conditions for path sensitization. To ensure the highest quality of path delay fault testing, each path delay should be tested under the most stringent sensitization criterion for which a test exists. Given various path sensitization criteria, paths are generally classified into several classes: single-path sensitizable, robust, non-robust, functional sensitizable and functional unsensitizable.

Some path delay faults do not need to be tested to guarantee the performance of the circuit. This is because these path delay faults can never independently affect the performance. There are many different ways to partition the set of paths into the set that needs to be tested and the set that does not need to be tested.

This chapter describes two criteria for classifying path delay faults. The first one is based on the path sensitization and the second is based on whether or not the given path needs to be tested to guarantee the performance of the circuit. Both single and multiple path delay faults are considered.

5.1 SENSITIZATION CRITERIA

Testing delay faults requires two vector patterns. Accordingly, path sensitization criteria are defined with respect to two vectors. This section addresses the sensitization of single path delay faults. Multiple path delay faults are addressed in Section 5.3.

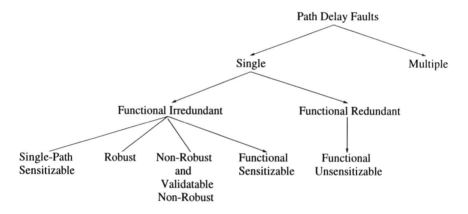

Figure 5.1. Path delay fault classification.

There exist several classes of path delay faults according to the sensitization criteria: single-path sensitizable, robust, non-robust, validatable non-robust, functional sensitizable and functional unsensitizable faults. These classes have different testability characteristics based on the specific fault detection conditions. The robust, non-robust, validatable non-robust and functional sensitizable faults can affect the performance of the circuit and they are together called **functional irredundant faults**. Functional unsensitizable faults, also called **functional redundant faults**, never independently determine the performance and they do not have to be tested. This section considers only functional irredundant faults, while functional unsensitizable faults will be analyzed in Section 5.2. The path delay fault classification used in this book is illustrated in Figure 5.1. Note that most of the literature on path delay faults considers the non-robust set as a superset of the robust set of paths and the functional sensitizable set as a superet of the non-robust testable set. However, in this book the set of robust testable, non-robust testable and functional sensitizable path delay faults are considered to be disjoint. The terminology used in the rest of the book is given next.

Terminology

An input to a gate is said to have a **controlling value** (denoted as cv) if it determines the value of the gate output regardless of the values on the other inputs to the gate. If the value on some input is the complement of the controlling value, the input is said to have a **non-controlling value** (denoted ncv). For example, in the case of an AND gate, the controlling value is 0 and the non-controlling value is 1, while for an OR gate the controlling value is 1 and the non-controlling value is 0. If the value of an input changes from the controlling to the non-controlling value, then the transition is denoted as cv→ncv. The ncv→cv transition changes the input from ncv to cv value. Each path delay fault is associated with a path and terms "path" and "path delay fault" will be used interchangeably. A signal is an **on-input** of path P if it is on P. A signal is an **off-input** of path P if it is an input to a gate in P but it is not an on-input. A path P is said to be **static sensitizable** if there is at least one input vector v such that all off-inputs in P settle at respective non-controlling values under vector v. A path is said to be a **false path** if it can never propagate a transition to the primary output. The logic functions computed by the gates and their propagation delays preclude false paths from being sensitized [42]. Paths that are not false are called **true paths**.

5.1.1 Single-path sensitizable path delay faults

Single-path sensitization criterion guarantees that the design under test will fail if and only if the path under test has excessive delay. This fault model fully characterizes the timing of the target path and it is ideal for delay fault diagnosis.

EXAMPLE 5.1 Consider an AND gate with two inputs, a and b, as shown in Figure 5.2. Let this gate be part of the target path P with input a as the on-input and input b as an off-input for P. Symbol S0 (S1) is used to denote a stable 0 (1) value on some signal under

Figure 5.2. Single-path propagation through an AND gate.

$V = \langle v_1, v_2 \rangle$. If the input a is assigned a rising or falling transition while input b is assigned a stable non-controlling value, the fault effect from the target path will always be observable at the output. Also, under these conditions, the fault effect at the output can only come from the target path and, therefore, this path delay fault is diagnosable.

Under the single-path sensitization criterion, all off-inputs in the target path have to be assigned stable non-controlling values. Since under this sensitization condition, the fault detection and diagnosis are guaranteed independent of circuit delays, tests for single-path sensitizable faults should be applied whenever they exist. However, single-path sensitization condition is very stringent and only a very small number of paths can be tested under this condition.

5.1.2 Robust testable path delay faults

The **robust sensitization criterion** [98, 139] allows unconditional detection of a path delay fault, while the sensitization conditions are less stringent then the single-path sensitization. In other words, if there is a fault on the target path that is sensitizable under robust sensitization criterion the fault will be observed independent of the delays on signals outside the target path.

Figure 5.3. Robust propagation through an AND gate.

EXAMPLE 5.2 Consider the AND gate in Figure 5.3. The values containing symbol X denote unspecified values under vector pair $V = \langle v_1, v_2 \rangle$,. For example, if the value on some signal is X1, it means that the value of the signal is unspecified for vector v_1 and it is 1 for vector v_2. If on-input a is assigned a rising transition and the off-input b is also assigned a rising transition, the fault effect from a will propagate to the output of the AND gate, whether or not there is a fault on the off-input b. This is because the output of the AND gate is determined by the later of the two rising transitions. Similarly, if the off-input b has a stable non-controlling value and the on-input a has either a falling or a rising transition, the fault effect from the target path will always be observable at the output. These sensitization conditions are called the robust sensitization conditions.

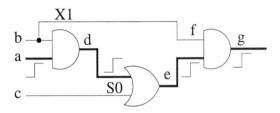

Figure 5.4. Robust path.

DEFINITION 5.1 Let f denote the on-input to gate g in the target path. Let h denote an off-input to gate g. The off-input h is called **robust off-input** with respect to an input vector pair V if:

(1) there is a cv→ncv transition or stable non-controlling value on h when the on-input f has a cv→ncv transition, and

(2) there is a non-controlling value on h when the on-input f has a ncv→cv transition.

DEFINITION 5.2 A path delay fault for which there exists an input vector pair such that it activates the required transitions on the path and all off-inputs in the path are robust off-inputs is called a **robust testable path delay fault**.

Therefore, a robust sensitizable path is static sensitizable.

EXAMPLE 5.3 Consider the circuit in Figure 5.4. We use symbols ↑ and ↓ to denote a rising and falling transitions, respectively. To specify a path delay fault, we list the transition at the source of the path and the signals in the path. For example, path delay fault $\{\uparrow, adeg\}$ represents the fault with the rising transition at the source of path consisting of signals a, d, e and g. This path is shown in bold in Figure 5.4. Path $\{\uparrow, adeg\}$ is a robust testable path because there exists a vector pair such that all off-inputs for this path are robust off-inputs.

For each robust testable path there could be more than one input vector pair that robustly sensitizes the path. However, to test a robust path it is sufficient to apply one such vector pair since the fault is guaranteed to be detected. For testing purpose, robust tests are highest quality tests for path delay faults and should be applied whenever they exist. However, experience shows that for most circuits only a small portion of path delay faults can be tested under the robust condition [32, 50]. Paths that cannot be tested under robust sensitization criterion are called **robust untestable path delay faults**. All robust testable paths have to be selected for testing to guarantee the performance of the circuit.

Figure 5.5.   Non-robust sensitization criterion for AND gate.

5.1.3   Non-robust testable path delay faults

The **non-robust sensitization criterion** [98] is less stringent than the robust criterion. This is because the detection of a fault on a path that is sensitizable under a non-robust criterion depends on the arrival times at the off-inputs and, thus, on the delays on certain signals outside the target path.

EXAMPLE 5.4 Figure 5.5 illustrates the non-robust propagation criterion for an AND gate. If there is a falling transition on the on-input a and a rising transition on the off-input b, the transitions on the output of the AND gate depend on the arrival times of the input transitions. If the transition on the off-input occurs later than that on the on-input, it will mask the propagation of the fault from the on-input to the output. In this case, the non-robust test is said to be *invalidated*. On the other hand, if the transition on the off-input occurs before the one on the on-input, the fault effect from the on-input will be observable at the output.

DEFINITION 5.3 Let f denote the on-input to gate g in the target path and h denote an off-input to gate g. The off-input h is called a **non-robust off-input** with respect to an input vector pair V if there is a ncv→cv transition on the on-input f and a cv→ncv transition on the off-input h under V.

DEFINITION 5.4 A robust untestable path delay fault for which there exists at least one vector pair such that it activates the required transitions on the path and at least one off-input is a non-robust off-input while the rest of the off-inputs are robust is called **non-robust testable path delay fault**.

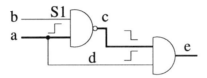

Figure 5.6. Non-robust path.

Therefore, non-robust testable paths are also static sensitizable.

EXAMPLE 5.5 Consider the circuit in Figure 5.6. Path $\{\uparrow, ace\}$ is a non-robust testable path and for the test shown in Figure 5.6 signal d is the non-robust off-input. If the rising transition on signal d arrives later than the transition on signal c, it will mask the propagation of the falling transition from signal c to signal e. In this case the test shown in the figure will not be able to detect the faulty target path (shown in bold).

There could be many different ways to non-robustly sensitize a given path, i.e., a non-robust testable path can have several possible non-robust tests. These non-robust tests differ in the number and positions of non-robust off-inputs in the given target path. A better non-robust test is the one for which the transitions on the non-robust off-inputs have a lower chance to mask the transitions on the target path. Finding such non-robust tests requires knowledge about the delays in the circuit. Therefore, including the timing information into the test generation process for non-robust paths can result in higher quality non-robust tests. A technique for generating such high quality non-robust tests is described in Chapter 7.

Paths that cannot be tested under robust or non-robust sensitization criteria are called **non-robust untestable paths**. All non-robust testable paths have to be selected for testing to guarantee the performance of the circuit.

5.1.4 Validatable non-robust testable path delay faults

Let a target path be $\{g_1, f_1, g_2, \ldots, g_n\}$, where g_1 and g_n denote a

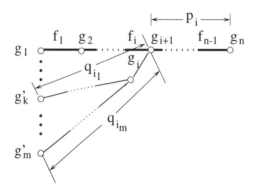

Figure 5.7. Validating a non-robust test.

primary input and a primary output, respectively, and f_i denotes the on-input feeding gate g_{i+1} (Figure 5.7). Suppose that under some non-robust input vector pair V for this path, signal h_i is a non-robust off-input for gate g_{i+1}. Let the partial path from g_{i+1} to g_n be p_i. Under a non-robust test the off-input h_i must be assigned a cv→ncv transition. This transition has to propagate to h_i through one or more partial paths from primary inputs under vector pair V. Let these paths be $q_{i_1}, \ldots q_{i_m}$. If all paths $q_{i_1} + p_i, \ldots, q_{i_m} + p_i$, where symbol "+" denotes path concatenation, can be robustly tested and if the circuit passes these tests, it can be guaranteed that the non-robust test for the target path will not be invalidated. The target path delay fault is called a **validatable non-robust path delay fault** [129]. The robust tests for the concatenated paths together with the non-robust test V for the target path form a validatable non-robust test. For example, consider again the circuit in Figure 5.6. Path $\{\uparrow, ade\}$ is the only path through signal d that can invalidate the non-robust test shown in the figure. However, this path is robust testable and can be checked for faults prior to applying the non-robust test. If it is faulty, the circuit will be classified as defective and applying the non-robust test is not necessary. If it is not faulty, the non-robust test in Figure 5.6 is guaranteed to detect the fault on the target path.

Not every non-robust testable path can be tested under validatable non-robust condition. However, a validatable non-robust test is the highest quality non-robust test and should be applied whenever it exists for testing non-robust testable paths. Automatic test generation for validatable non-robust tests is a complex problem. Some results have been reported in [141]. An algorithm for generating validatable non-robust tests will be presented in Chapter 7.

Similar to the robustly testable faults, many circuits have only a small percentage of non-robust testable faults.

5.1.5 Functional sensitizable path delay faults

As with the non-robust sensitization criterion, detection of faults on paths that are sensitizable under **functional sensitization criterion** [32] depends on the delays on signals outside the target path. The

Figure 5.8. Functional sensitizable propagation for AND gate.

functional sensitization criterion requires that there exists more than one faulty path in the circuit in order for the target fault to be detected.

EXAMPLE 5.6 Figure 5.8 illustrates the functional sensitizable criterion for an AND gate. When the on-input a and off-input b both have falling transitions, in order to propagate the fault to the output of the AND gate both transitions have to be late. This is because the arrival time of the signal at the output is determined by the earlier of the two falling transitions.

DEFINITION 5.5 Let a signal f be the on-input to gate g in a target path. Let signal f be the on-input to gate g. The off-input h is called **functional sensitizable (FS) off-input** with respect to an input vector pair V if there is a ncv→cv transition on both on-input f and the off-input h.

DEFINITION 5.6 A non-robust untestable path delay fault for which there exists an input vector pair such that it activates the required transitions on the path and at least one off-input is FS off-input while the rest of the off-inputs are either robust or non-robust is called **functional sensitizable path**.

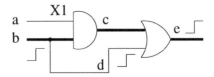

Figure 5.9. Functional sensitizable path.

A functional sensitizable path is not static sensitizable. If for a given input vector pair that functionally sensitizes some FS path, all of its FS off-inputs are late, the FS path might determine the performance of the circuit. However, if at least one FS off-input is not late, the FS path cannot impact the performance, i.e., it is a false path.

EXAMPLE 5.7 Consider the circuit in Figure 5.9. Let the target path be $\{\uparrow, bce\}$. This path is functional sensitizable because under any input vector pair V that propagates a rising transition on this path, the off-input d must be assigned ncv→cv transition (d is an FS off-input).

For each FS off-input there must exist one or more partial paths from primary inputs through which ncv→cv transition can reach the FS off-input. These paths are said to be *associated* with the given off-input.

EXAMPLE 5.8 In Figure 5.9 partial path $\{\uparrow, bd\}$ is associated with the FS off-input d. The value of signal e in the target path is determined by the input that arrives earlier. Therefore, in order to detect the fault on the target FS path, the transition on the FS off-input d must also be late. This means that the fault effect on the target path can only be observed in the presence of multiple path delay faults.

An FS path can be tested with several different tests. Since the detection of a fault on an FS path depends on the delays on certain other signals in the circuit, for the same target path, different FS tests have a different probability of detecting the defect. In Chapter 7 we describe a methodology for generating good quality FS tests using the circuit timing information.

Functional sensitizable paths can affect the performance only if groups of FS paths are simultaneously faulty. These groups of FS paths belong to the class of faults called **primitive faults** [78, 79]. The primitive faults will be described in greater detail in Section 5.3. A given FS path can belong to many primitive faults. All these primitive faults have to be tested to guarantee that a given FS path delay fault will not affect the performance of the circuit. Therefore, a number of different FS tests have to be applied to test a functional sensitizable path.

Some functional sensitizable paths do not have to be tested to ensure the temporal correctness of the circuit. The properties and identification of such FS paths are discussed in the next section.

5.2 PATH DELAY FAULTS THAT DO NOT NEED TESTING

The main disadvantage of the path delay fault model is the large number of paths in the circuit. For this reason test generation and synthesis for path delay fault testability usually cannot be done for large designs using reasonable amount of resources. Also, a large number of faults can imply a large test set and long test application time.

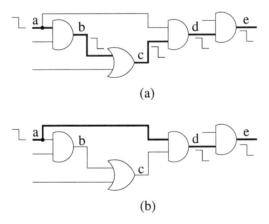

Figure 5.10. Example of a path that does not have to be tested.

Recent research [32, 53, 92, 140] shows that it is usually not necessary to test all path delay faults to guarantee the circuit performance. This is because there exist path delay faults that can never impact the circuit performance unless some other paths also have delay faults. These paths do not have to be tested if the other paths have been tested.

EXAMPLE 5.9 Consider the circuit in Figure 5.10(a). Path P_1 (shown in bold lines) consists of gates a, b, c, d and e. Let P_2 be the path consisting of gates a, d and e (shown in bold lines in Figure 5.10(b)). Clearly, if path P_2 does not have a delay defect that slows down the propagation of a falling transition, then the value on gate d is determined by signal a and not by signal c. Therefore, delay defects on path P_1 can affect the delay of the circuit only if path P_2 also has a delay defect. This implies that path P_1 for a falling transition does not have to be tested if path P_2 for falling transition is tested.

There are many different ways to partition the set of all path delay faults into the set that needs to be tested and the set that does not need to be tested. The test set size and the test generation effort depend on the number of faults in the set that needs to be tested. Therefore, it is important to find a set with a minimum number of faults. However, identifying such a minimum set of paths that need to be tested is a complex problem. This section describes several methods that have been proposed for partitioning the set of path delay faults. The classification into functional irredundant and functional redundant path delay faults proposed by Cheng and Chen [32] is described first. After that, the classification into robust testable and robust dependent path delay faults proposed by Lam et al. [92] is outlined. Next, the methods proposed by Sparmann et al. [140] and by Gharaybeh et al. [55] for identifying the paths that do not need to be tested are given. Finally, the classification into primitive and non-primitive path delay faults first proposed by Ke and Menon [78] is described. Given the path delay fault classification in Figure 5.1, these methods agree that all robust, non-robust and validatable non-robust path delay faults have to be selected for testing to guarantee the circuit performance. They also agree that none of the functional unsensitizable faults need to be tested since they can never independently determine the critical path delay for the circuit. The difference in these techniques is in partitioning the set of FS paths. Many designs have a very large percentage of functional sensitizable paths and finding a minimal set of the FS paths that have to be tested represents an important problem. Comparison of the experimental results obtained using the above techniques shows the trade-offs between the speed and the accuracy of the proposed algorithms.

5.2.1 Functional irredundant vs. functional redundant path delay faults

Cheng and Chen [32] propose dividing all paths into two sets: one that contains all robust, non-robust and functional sensitizable paths and the other that contains functional redundant (also called functional unsensitizable) paths. Paths in the first set determine the performance of the circuit while the functional unsensitizable paths can never influence the performance of the circuit. **Functional unsensitizable path delay faults** are defined as the faults for which under all possible input vector pairs, either (1) at least one off-input in the path has a controlling value

Figure 5.11. Functional unsensitizable off-inputs for an AND gate.

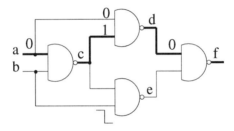

Figure 5.12. Functional unsensitizable path.

under vector v_2 when the corresponding on-input has a non-controlling value, or (2) at least one off-input is assigned a stable controlling value. Such off-inputs are called **functional unsensitizable off-inputs**.

EXAMPLE 5.10 Figure 5.11 illustrates functional unsensitizable off-inputs for an AND gate. As an example of a functional unsensitizable path consider the circuit in Figure 5.12. The values in the figure are the values for the second input vector of the vector pair. Path $\{\downarrow, acdf\}$ is a functional unsensitizable path because the propagation of the transition on the target path is stopped at gate d due to the assignment of a controlling value at the off-input and a non-controlling value at the on-input.

To find the set of functional unsensitizable paths Cheng and Chen [32] identify a set of non-overlapping, functional unsensitizable prime segments.

58   CHAPTER 5

DEFINITION 5.7 A partial path $Q = \{f_s, g_{s+1}, f_{s+1}, \ldots, f_{t-1}, g_t\}$ is called a **functional unsensitizable zero (one) segment** if there exists no input vector such that:

(1) it stabilizes f_s at logic 0 (logic 1), and

(2) for each gate g_{i+1} in Q whose on-input is stabilized at a non-controlling value, it stabilizes all off-inputs at their respective non-controlling values.

Any path that covers a functional unsensitizable segment is a functional unsensitizable path. Identification of functional unsensitizable prime segments (defined below) allows the counting of the number of functional unsensitizable path delay faults in a given circuit.

DEFINITION 5.8 A functional unsensitizable zero (one) segment Q is a **prime segment** if no proper sub-segment of Q is a functional unsensitizable zero (one) segment.

The algorithm for identifying functional unsensitizable paths has two phases [32]. The first phase finds a partial path $Q = \{g_0, f_0, g_1, \ldots, f_{t-1}, ig_t\}$, where g_0 is a primary input, f_i is the on-input to gate g_{i+1} and g_t is the destination of the partial path Q, such that Q is a functional unsensitizable segment. The second phase locates gate g_{s+1} in Q closest to g_t such that $S = \{f_s, g_{s+1}, \ldots, f_{t-1}, g_t\}$ is a functional unsensitizable prime segment. In both phases, the partial path is expanded forward connection by connection (gate by gate). In the beginning, the partial path Q is initialized as $\{g_0, f_0, g_1\}$. The primary input g_0 is assigned value 0 or 1. If signal f_0 is set to a non-controlling value, it is required that all off-inputs of Q are also set to non-controlling values. If f_0 is set to controlling value, there are no requirements on the off-inputs of Q. If the current requirements for Q cannot be satisfied, partial path Q is a functional unsensitizable segment, and the first phase stops. Otherwise, the partial path Q is expanded forward to include the next pair consisting of a signal and a gate. Again, if the newly included signal is set to its non-controlling value, all off-inputs to the newly included gate must be set to the non-controlling value as well. If the expanded partial path Q is not a functional unsensitizable segment, the procedure continues.

Let the final partial path after the first phase be $Q = \{g_0, f_0, g_1, \ldots, f_{t-1}, g_t\}$. In the second phase, partial path S is initialized to be $\{f_{t-1}, g_t\}$. If on-input f_{t-1} is assigned a non-controlling value, all off-inputs for gate

g_t must be assigned non-controlling values. There are no requirements on off-inputs, if the on-input f_{t-1} is assigned a controlling value. If S is determined to be functional unsensitizable segment, a prime segment closest to g_0 is located and the second phase stops. Otherwise, the partial path S is expanded backwards and the procedure repeats until S is a prime segment. Once a prime segment is identified, the algorithm continues to search for the next segment until the search space is exhausted. Suppose the last segment generated in the first phase is $Q = \{g_0, f_0, g_1, \ldots, f_{t-1}, g_t\}$. The new Q is formed by removing the last pair of signal and gate, $\{f_{t-1}, g_t\}$, and by adding $\{f'_{t-1}, g'_t\}$, where f'_{t-1} represents the next fanout of gate g_{t-1} and a fanin to gate g'_t. If g_{t-1} does not have any unexplored fanouts, $\{f_{t-2}, g_{t-1}\}$ is removed and g_{t-2}'s next fanout is added to Q, etc. To determine if a path is functional unsensitizable, the value requirements on off-inputs must be justified. Since a complete justification process is very time consuming for large circuits, to make the procedure efficient only mandatory assignments [81] and their implications are used to find paths that are functional unsensitizable.

The identified set of paths that does not have to be tested using the above procedure can be suboptimal (too large) for two reasons. First, only the second vector of the vector pair required to launch a transition is considered to identify functional unsensitizable paths. This means that some paths that under all possible input vector pairs have a stable controlling value on some off-input are not classified as functional unsensitizable even though they can never affect the performance of the circuit. Second, only local implications are used to identify functional unsensitizable faults. Fast algorithms for identification of functional unsensitizable faults based on computing static logic implications were recently proposed by Li et al. [96] and Heragu et al. [66].

5.2.2 Robust vs. robust dependent path delay faults

Lam et al. [92] partition all path delay faults into two disjoint sets: the set of robust testable delay faults and the set of **robust dependent (RD) faults**. Robust dependent faults are faults that cannot impact the performance of the circuit unless a fault also occurs on some robust testable path.

DEFINITION 5.9 [92] Let $\mathcal{D}$ be the set of all path delay faults in circuit C and $\mathcal{R}$ a subet of $\mathcal{D}$. If for all τ, where τ is the delay of the circuit, the absence of delay faults in $\mathcal{D} - \mathcal{R}$ implies that the delay of C is smaller or equal τ, $\mathcal{R}$ is said to be **robust dependent (RD)**.

The RD path-set is independent of delays of the signals, i.e., faults in the RD set can be eliminated from testing regardless of the delay values of the signals in the circuit.

The method in [92] finds an RD set given that the circuit is represented as a leaf-dag. **Directed acyclic graph (dag)** is a directed graph with no directed cycles [136]. A **leaf-dag** i represents a rooted dag in which paths that start from the root reconverge only at the input vertices (leaves) [36]. Therefore, a leaf-dag represents a circuit consisting of AND and OR gates with multiple fanouts and inverters permitted only at the primary inputs, and with each inverter allowed only a single fanout. Every circuit can be represented as a leaf-dag by gate duplication. There is a one-to-one correspondence between the paths in a circuit and its leaf-dag. The **I-edge** of a path in a circuit is either a connection from the primary input if no inverter is present or the connection immediately after the inverter. A **falling (rising) RD path-set** represents a set of RD path delay faults with falling (rising) transitions at the primary outputs. The following theorem gives sufficient conditions under which a set of paths is an RD path-set.

THEOREM 5.1 [92] Let C be a given circuit and η its leaf-dag. Let $\mathcal{R}_\eta$ be the paths in η corresponding to the set of paths $\mathcal{R}$ in C. Let $\mathcal{M}_\eta$ be the I-edges of $\mathcal{R}_\eta$. If multiple stuck-at-0 (stuck-at-1) fault on $\mathcal{M}_\eta$ is redundant in η, then $\mathcal{R}$ is a rising(falling) RD path-set in C.

The above theorem links the identification of RD path-set for circuit C to finding redundant multiple stuck-at faults in a leaf-dag of C. In practice, due to high CPU-time and memory requirements, identification of redundant multiple stuck-at faults and the transformation of the circuit to a leaf-dag are not easy to perform. Therefore, Lam *et al.* [92] propose an approximate technique to find an RD path-set. It is based on identifying redundant multiple stuck-at faults by iteratively identifying redundant single stuck-at faults. It also eliminates the need to completely unfold the given circuit into a leaf-dag. The algorithm, which finds a maximal RD path-set (with no claims on how close it is to the maximum RD path-set), operates on an internally noninverting circuit. **Internally noninverting circuit** is a circuit that has inverters only at the primary inputs. The algorithm is based on the following theorem:

THEOREM 5.2 [92] Let C be an internally noninverting circuit and M be a redundant multiple stuck-at-0 (stuck-at-1) fault in C (faults considered on or after I-edges). Let C_M be the circuit obtained by replacing each connection in M by 0 (1). If P is rising (falling) robust path delay fault in C, then P is a rising (falling) path delay fault in C_M.

An internally noninverting circuit C' can be obtained for a given circuit C by duplicating gates in C. C' is at most twice the size of C. Identification of the RD path-set is done by applying the following two steps. First, replace each redundant stuck-at-0 (stuck-at-1) connection by a 0 (1) to obtain an irredundant circuit. Second, duplicate selected gates to obtain a fanout-free circuit. Since in the second step, new redundancies might be created, the two steps are iterated until the resulting circuit is fanout-free and irredundant. The paths in the resulting circuit that do not pass through any constant connection form the non-RD set.

This procedure identifies a near maximum RD set. However, it is very time and space consuming and can be applied only to small scale circuits. The procedure described in the previous section is very efficient and can be applied to much larger circuits. However, the identified paths that do not have to be tested form a superset of the RD path-set.

5.2.3 Path classification based on input sort heuristic

Sparmann et al. [140] propose a technique that shows a trade-off between the efficiency and the size of the identified set of faults that do not require testing. This technique combines the method described in Section 5.2.1 and a heuristic that implicitly orders the paths to classify them into the set that has to be tested and the set that does not need to be tested. In comparison with the method in Section 5.2.1, this method can be applied to larger designs and it identifies a larger set of paths that do not have to be tested. Like the method of Section 5.2.1, this procedure also relies only on mandatory assignments and their implications to find the set of paths that do not require testing. It also uses only the second vector of the vector pair to perform the classification but in addition to identifying functional unsensitizable paths, it also identifies some functional sensitizable paths that do not have to be tested. Functional sensitizable faults must occur together with some other faults in order to affect the performance. As will be explained in Section 5.3, for each such group of simultaneously faulty paths it is sufficient to select one functional sensitizable path for testing. To find a minimal path cover for testing, the heuristic in [140] orders the inputs of each gate in the circuit. This order of inputs is called an **input sort**.

Given a target path and an input sort, the on-input partitions the off-inputs into higher and lower order off-inputs. For example, consider the

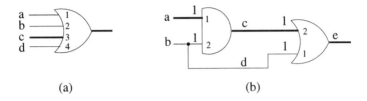

Figure 5.13. Using the *input sort* heuristic.

four-input OR gate of Figure 5.13(a). Inputs a, b, c and d are assigned integers 1, 2, 3 and 4, respectively. If input c is the on-input, then (for any target path through c) inputs a and b are lower order off-inputs while d is a higher order off-input. Given a path delay fault and an input sort, if a lower order off-input cannot be assigned a non-controlling value, then the off-input is **partially static unsensitizable (PSUS)**. It can be shown that if an FS path has at least one PSUS off-input for all possible input vectors, then the FS path does not have to be tested [140]. These paths will not affect the performance of the circuit unless some robust, non-robust or other functional sensitizable path is faulty at the same time. For example, consider the circuit shown in Figure 5.13(b). The path consisting of gates a, c and e is functional sensitizable for a rising transition. However, the off-input d has a lower order than on-input c and d always assumes a controlling value. Therefore, the target path does not have to be tested. Defects on this path cannot affect the performance unless path $\{\uparrow, bde\}$ is also defective.

Different input sorts can lead to different partitions of paths into the "to be tested" and "need not be tested" sets. A "good" input sort results in a smaller set of paths that have to be selected for testing. Several different heuristics for finding a "good" input sort have been proposed [140].

5.2.4 Path classification based on single stuck-at fault tests

Gharaybeh *et al.* [53] derive a logic model for delay faults and show that path delay faults can be classified using single stuck-at fault test generation process on this model. The set of path delay faults is classified into three categories: singly-testable, multiply-testable and singly-testable dependent. Singly- and multiply-testable paths must be tested to guarantee the performance of the circuit, while singly-testable dependent paths do not have to be tested. **Singly-testable path delay faults** are those that can be guaranteed to be detected under the assumption that no other delay fault exists in the circuit. These faults are either robust, non-robust or validatable non-robust. **Singly-testable dependent path delay faults** cannot affect the performance unless a fault simultaneously happens on some singly-testable path. This set is a superset of the functional unsensitizable set of paths [32] and a subset of the RD set [92]. In addition to functional unsensitizable faults it also contains some of the functional sensitizable faults that do not have to be tested. **Multiply-testable path delay faults** are faults that can affect the performance only together with some other paths and none of these paths is singly-testable dependent. This set contains the functional sensitizable faults that are not in the singly-testable dependent faults set.

5.2.5 Primitive vs. non-primitive path delay faults

The classification of path delay faults into the set of primitive and set of non-primitive faults was first proposed by Ke and Menon [78]. Later a similar classification was proposed by Krstić et al. [87] and Sivaraman and Strojwas [137]. *Primitive faults* [78, 79] represent faults that have to

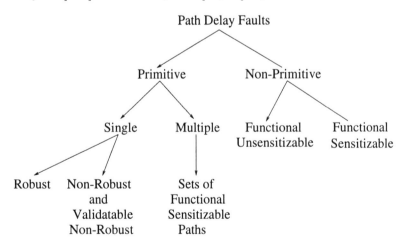

Figure 5.14. Primitive faults vs. non-primitive faults.

be tested in order to guarantee the temporal correctness of the circuit. On the other hand, testing of *non-primitive faults* is not required if tests for primitive faults have been derived. This is because non-primitive faults can never independently affect the performance of the circuit. Primitive faults can be single or multiple. Single primitive faults usually represent a small portion of all primitive faults. Most of the primitive

64   CHAPTER 5

faults consist of more than one faulty path. Figure 5.14 illustrates the classification of path delay faults into primitive and non-primitive faults. Primitive faults will be discussed in detail in the next section.

5.3   MULTIPLE PATH DELAY FAULTS AND PRIMITIVE FAULTS

The following terminology, taken from [78], is needed in order to formally define primitive faults. A **multiple path** Π is a set of single paths

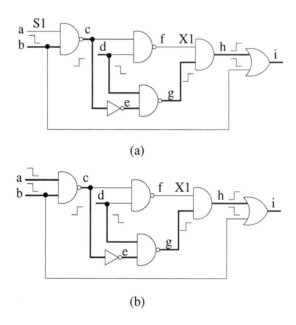

Figure 5.15.   Primitive faults.

that end at the same primary output. A **multiple path delay fault** (MPDF) on Π is a condition under which every single path in Π has a fault. A signal is called an **on-input of a multiple path** Π if it is on Π. Therefore, a multiple path delay fault can have a gate such that several of its inputs are on-inputs (**multiple on-input**). A signal is called an **off-input of a multiple path** Π if it is an input to a gate in Π but it is not an on-input. An MPDF on π is said to be static sensitizable if there exists at least one input vector pair $V = \langle v_1, v_2 \rangle$ such that it launches the required transitions on all primary inputs in Π and each off-input of Π assumes a final non-controlling value.

EXAMPLE 5.11 In the circuit in Figure 5.15, single path delay faults $P_1 = \{\downarrow, bceghi\}$ and $P_2 = \{\downarrow, dghi\}$ form a double path delay fault $\Pi = \{P_1, P_2\}$. Gate g has a multiple on-input consisting of two signals, d and e. This MPDF is static sensitizable.

DEFINITION 5.10 [78] A **primitive fault** is defined as a multiple path delay fault that satisfies the following two conditions:

(1) the MPDF is static sensitizable, and

(2) no proper subset of this MPDF is static sensitizable.

The number of single path delay faults contained in a primitive fault represents the **cardinality** of a primitive fault. The robust and non-robust path delay faults satisfy the above two conditions and can be regarded as primitive faults of cardinality 1. Since the functional sensitizable faults are not static sensitizable they do not satisfy the first condition for a primitive fault. However, several FS paths together can form a primitive fault. In fact, from the second condition required for primitive faults it follows that all single paths in a primitive fault of cardinality higher than 1 must be functional sensitizable paths.

EXAMPLE 5.12 Consider again the circuit in Figure 5.15. Faults $P_1 = \{\downarrow, bceghi\}$ and $P_2 = \{\downarrow, dghi\}$ are FS faults. When considered independently, none of these two faults is static sensitizable because it is impossible to find an input vector pair, $V = \langle v_1, v_2 \rangle$, such that all off-inputs to this multiple delay fault are assigned non-controlling values under vector v_2. However, the double path delay fault $\Pi_1 = \{P_1, P_2\}$ is static sensitizable. Gate g has a multiple on-input and no off-inputs. Therefore, MPDF Π_1 is a primitive fault of cardinality 2. If both single paths (P_1 and P_2) are faulty, the test for Π_1 will make the fault observable at the primary output.

Figure 5.14 illustrates the relationship between different classes of single path delay faults and primitive faults.

Since all single paths in a primitive fault end at the same primary output, every primitive fault of cardinality higher than 1 must have at least one gate with a multiple on-input. Such gates are called **merging gates** of the primitive fault [84, 87]. In the above example, gate g is a merging gate for primitive fault $\Pi_1 = \{P_1, P_2\}$. All signals in a multiple on-input must be assigned ncv→cv transition. Single paths in a primitive fault are said to be **co-sensitized** at the merging gates.

EXAMPLE 5.13 Consider the circuit in Figure 5.15(b). Fault $P_3 = \{\downarrow, aceghi\}$ is also an FS fault. The test shown in the figure sensitizes three FS paths, $\Pi_2 = \{P_1, P_2, P_3\}$ and they merge at gates c and g. However, Π_2 is not a primitive fault because it does not satisfy the second condition for primitive faults (since $\Pi_1 \subset \Pi_2$). To understand this condition, assume that the circuit has passed the test for Π_1. It means that at least one of the single paths (P_1 or P_2) is not faulty. Let path P_1 be faulty and path P_2 be delay fault free. The test for Π_2 cannot detect the fault on P_2 because it requires that paths P_1 and P_3 are faulty together with P_2. Therefore, if the test for Π_1 cannot detect the fault in the circuit, neither can the test for Π_2. Next, consider that the circuit did not pass the test for Π_1. In that case, the circuit will be classified as delay-defective and no testing of Π_2 is required.

Identification and testing for primitive faults is a complex problem. The testing strategies for the primitive faults will be addressed in Chapter 7.

5.4 PATH DELAY FAULT CLASSIFICATION FOR SEQUENTIAL CIRCUITS

Identifying path delay faults that do not have to be tested in non-scan sequential designs is important because the number of sequential paths, often unknown, is significantly larger than the number of combinational paths (paths in the combinational part of the sequential design). In general, more paths can become testable if a slow-fast-slow clock is used instead of the rated-clock [103]. The discussion here is related to path testability with rated-clock. A **sequential path** is a concatenation of several combinational paths in the iterative array model of the sequential circuit (Figure 2.2). Tekumalla and Menon [142] propose a method for identifying faults that do not have to be tested in sequential circuits for which state transition diagrams are available. However, constructing state transition diagrams for most practical circuits is intractable.

In combinational circuits, path delay faults can be classified into two disjoint sets, as shown in Figure 5.16.

DEFINITION 5.11 **Path delay fault cover (PDFC)** is the set of faults that should be considered for delay testing.

PATH DELAY FAULT CLASSIFICATION   67

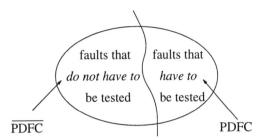

Figure 5.16.   Fault classification in combinational circuits.

For combinational circuits the PDFC is identical to the set of primitive faults described in the previous section. A method to identify PDFC for sequential circuits described as multi-level netlists has been proposed by Krstić et al. [83]. The faults in the PDFC of the sequential circuit either (1) cannot independently affect the performance of the circuit, or (2) have no test obtainable from gate-level test generation tools. The latter types are the **untestable path delay faults**. A proposed fault classification for sequential circuits is illustrated in Figure 5.17. Only the faults from the testable PDFC have to be considered for delay testing. In the case of sequential designs, the set of primitive faults contains all faults in the testable PDFC class plus some faults in the untestable faults class.

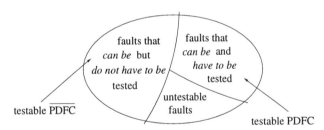

Figure 5.17.   Fault classification in sequential circuits.

The untestable path delay faults may adversely affect the delay of the circuit but gate-level delay test generators will be unable to find tests for these faults. Prior identification of these faults is desirable since test generators expend significant amount of computing resources on these faults. The method [83] for identifying faults that do not have to be considered for testing is based on two key ideas: (1) the use of the knowledge of the sequential behavior of the circuit to weed out faults that may belong to the PDFC of the combinational part but do not have to be included in the PDFC of the sequential circuit, and (2) classification of faults by considering both vectors that are required to launch a given transition on the target path. This is unlike most other classification techniques that ignore the first vector required to launch the transition because that usually entails high computational complexity.

5.4.1 Sequential PDFC

Segments are used to precisely define a sequential path. A **segment** is an ordered set of gates $g_0, g_1, \ldots, g_n$. Here, g_0 is a primary input or the output of a flip-flop, and g_n is a primary output or the input of a flip-flop. Also, gate g_i is an input to gate g_{i+1} ($0 \leq i \leq n-1$) and gates $g_1, \ldots, g_n$ are all Boolean logic gates.

EXAMPLE 5.14 Consider the circuit shown in Figure 5.18(a). The

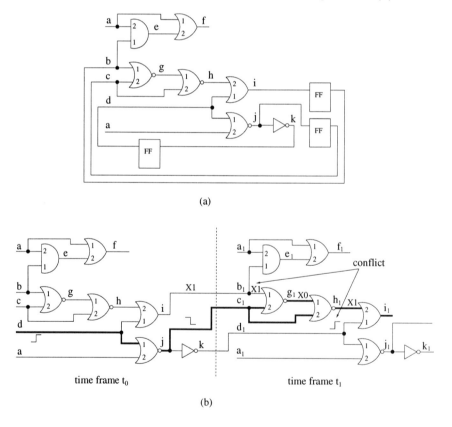

Figure 5.18. An example showing that PDFC_S ⊆ PDFC_C.

ordered set of gates d, j and k is a segment. This segment begins at flip-flop d and terminates at the gate k. A sequential path is a concatenation of segments. The first segment begins at a primary input and the last segment terminates at a primary output. A sequential path of three segments in the circuit in Figure 5.18(a) is as follows: aj, $cghi$ and bef. Note that the first segment begins at the primary input a. This segment terminates at the input of flip-flop c. The second segment begins at flip-flop c and terminates at the input of flip-flop b. The last segment terminates at the primary output f.

A sequential circuit can have a huge number of sequential paths. Also, unlike combinational circuits where the number of paths is known, it may not be possible to count the number of sequential paths in circuits with feedback cycles. Therefore, finding a path delay fault cover by considering one sequential path at a time may be computationally infeasible for circuits with feedback cycles. Instead, all sequential paths that include a given segment are simultaneously examined. Two faults are modeled for every segment to capture the propagation of the rising and falling transition through the segment. In the sequel, a path delay fault for a sequential circuit will be referred to as a **segment fault** and it is specified by a segment and a transition. There are two advantages of using segments. First, the target fault list for path delay fault testing has a finite number of faults. The set of modeled faults is equal to the number of path delay faults in the combinational logic. Second, a PDFC of the sequential circuit can be specified as a finite set of segments rather than a possibly infinite set of sequential paths. The segment delay fault model has been discussed in the literature and was defined in Section 3.5 [64].

Krstić et al. [83] distinguish between the PDFC of the combinational logic and the PDFC of the sequential logic. A PDFC of the combinational logic, denoted as PDFC_C, is derived assuming that all flip-flop output and input signals are also primary inputs and outputs, respectively. A segment fault that is in PDFC_C may not belong to the PDFC of the sequential circuit (denoted as PDFC_S). This happens when none of the sequential paths that include the segment can independently affect the delay of the circuit.

EXAMPLE 5.15 Consider the sequential circuit in Figure 5.18(a). The segment fault $\{\uparrow, dj\}$ is in PDFC_C because the fault is static sensitizable. However, this fault does not have to be included in PDFC_S because there is no sequential path through the segment dj that can propagate a rising transition and independently affect the circuit delay. To see this, consider two successive frames of the iterative array model of the circuit. These frames are shown in Figure 5.18(b). This figure also shows the input sort (see Section 5.2.3) for both frames. The implications of having a rising transition on signal d are indicated in the figure. Sequential paths through the segment dj will have to include either segment $c_1 g_1 h_1 i_1$ or the segment $c_1 h_1 i_1$ in the time frame t_1. The segment $\{\downarrow, c_1 g_1 h_1 i_1\}$ is functional unsensitizable. This is because the on-input c_1 has a non-controlling value and the off-input b_1 has a controlling value for the second vector of the vector pair that launches the transition on signal d. Therefore, no sequential path through segment $c_1 g_1 h_1 i_1$ has to be included in the PDFC of the sequential circuit. The segment $\{\downarrow, c_1 h_1 i_1\}$ is functional sensitizable. However, because of the input sort, gate i_1 has a partially static unsensitizable (PSUS) off-input d (defined in Section 5.2.3) for any test sequence that launches a rising transition on gate d. The off-input d_1 has a lower order and it assumes a controlling value for the second vector of the vector pair that launches the transition on d. Therefore, no sequential path through segment $c_1 h_1 i_1$ has to be included in the PDFC of the sequential circuit. This implies that we can exclude the segment fault $\{\uparrow, dj\}$ from PDFC_S.

In general, a sequential path does not have to be considered for delay testing if it has a functional unsensitizable (FUS) or a PSUS off-input for all input sequences that initiate a transition on the path. If a segment fault is not in PDFC_C, then there is already at least one FUS or PSUS off-input. Therefore, this fault does not have to be included in PDFC_S. Only a subset of faults in PDFC_C will be included in PDFC_S.

5.4.2 Untestable segment faults

Sequential circuits can have segment faults for which no test can be found by the test generator. Also, for some faults it may be impossible to prove that no test is possible from any initial state of the circuit.

EXAMPLE 5.16 Consider again the circuit in Figure 5.18(a) and the segment fault $\{\uparrow, chi\}$. It can be shown that this fault is static sensitizable if only the combinational logic is considered. Therefore, the fault is included in PDFC_C. If the sequential circuit is considered, the flip-flop c has to assume the value 1 to launch a transition on the segment. This implies that signals a and d have to assume the value 0. However, a test generator that starts with flip-flops in an unknown state will not be able to initialize d to the logic value 0. This is because d can be set to 0 during a clock period only if its value was 0 in the previous clock period. Since the test generator assumes that d starts with an unknown value, it will not be able to initialize d to 0. Therefore, it is not possible to determine if the segment fault $\{\uparrow, chi\}$ should be included in or excluded from PDFC_S. Such faults are referred to as **untestable segment faults**.

A delay test generator that processes one sequential path at a time will consider every sequential path through the segment and prove the path to be untestable. However, this method will require significant computing resources. It is desirable to identify untestable segments and eliminate them from consideration by a delay test generator. To reduce the number of untestable faults, one can consider each initial state of the circuit separately and determine if the sequential path has a test. If a delay test is possible for every initial state, then we can include the fault in PDFC_S. However, this method is impractical for most circuits of interest. Multiple observation time strategy [120] is another option but this technique also requires prohibitively high computational resources.

Several known techniques [7, 29, 97, 123] can be used to determine the set of values that a signal in the sequential circuit can or cannot assume in any time-frame. This information is referred to as the **functional signal constraint (FSC)** of a signal. The FSC information for a signal is derived assuming an unknown initial state for the sequential circuit. If a signal assumes a value of 0 (1) in every time frame, then we assign the symbol $C0$ ($C1$) to the signal. If a signal cannot assume a value 0 (1) in any time frame, then we assign the symbol $U0$ ($U1$). If it is impossible to justify a logic value of 0 or 1 on a signal, then we assign the symbol U. Finally, if both 0 and 1 values on a signal can be justified, then we assign the symbol G. These symbols, referred to as *FSC values*, have been used by Liang *et al.* [97].

EXAMPLE 5.17 Consider again the circuit in Figure 5.18(a). Since signal d cannot assume the value 0, it is assigned an FSC value of $U0$.

The FSC values are useful in quickly identifying untestable segment faults. If any signal on a target segment fault has an FSC value other than G, then it will not be possible to initiate a transition on the segment. Therefore, the segment fault is untestable. These faults are referred to as **unexcitable segment faults**.

EXAMPLE 5.18 Consider the AND gate shown in Figure 5.19(a). This gate has two inputs a and b. Let a and b be the on-input and

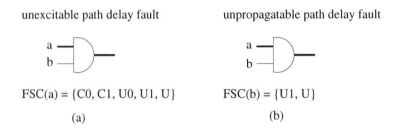

Figure 5.19. Rules for identifying untestable path delay faults through an AND gate.

off-input, respectively. If the FSC value of signal a is $C0$, $C1$, $U0$, $U1$ or U, then no transition can be launched on signal a. Therefore, all sequential paths through signal a are untestable. Note that this analysis considers both vectors of the vector pair that can initiate a transition on signal a. Based on the FSC value, it may be the first or the second vector of the transition initiating vector pair that cannot be justified starting from an unknown initial state of the sequential circuit.

If an off-input cannot assume a non-controlling value, then the test generator will be unable to derive a delay test. For example, consider the AND gate in Figure 5.19(b). If signal b has an FSC value of U or $U1$ (cannot assume a non-controlling value 1), then it will not be possible to derive a delay test for a sequential path through a. Such untestable segment faults are referred to as **unpropagatable segment faults**.

EXAMPLE 5.19 Figure 5.20 shows the FSC values for the circuit shown in Figure 5.18(a). The FSC values were derived using the symbolic simulation technique described by Liang et al. [97]. Only segments af and aef have to be considered by the delay test generator. All other segments correspond to untestable segment faults because a transition cannot be initiated on one or more on-inputs.

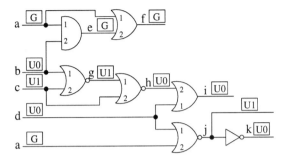

Figure 5.20.   FSC values for the circuit shown in Figure 5.18(a).

5.4.3 Algorithm for identifying testable PDFC for sequential circuits

The algorithm [83] for identifying testable PDFC for sequential circuits consists of two parts. First, a segment fault that belongs to the PDFC of the combinational logic is identified. The techniques described in Sections 5.2.1 and 5.2.3 can be used to identify segment faults in PDFC_C. However, these techniques do not take advantage of the sequential behavior of the circuit. Krstić et al. modify the algorithm of Section 5.2.3 to include the untestable segment fault identification ideas discussed earlier.

Second, a check is performed to see if the segment fault has to be included in the PDFC of the sequential circuit. Based upon the characteristics of the segment, one of the following steps is used to determine if the fault belongs to PDFC_S:

(1) If the segment starts at a PI and terminates at a PO, then there is only one sequential path through this segment. The segment fault is included in PDFC_S.

(2) If the segment starts at a PI and terminates as an input to a flip-flop, the *Forward* phase is entered. A finite number of frames in the iterative array model of the circuit is considered. The frames are labeled $t_0, t_1, \ldots, t_k$. Here, frame t_{i+1} immediately follows frame t_i ($0 \leq i \leq k-1$) in time. The target segment is in frame t_0. All paths in the iterative array model that include the target segment are implicitly examined. The sequential nature of the circuit is considered by implying mandatory assignments across time-frames. The iterative array model can be considered as a combinational circuit. Inputs to frame t_0 are considered as primary inputs and outputs of frame t_k

are considered as primary outputs. If no combinational path that includes the target segment has to be included in the PDFC of the iterative array model, then the segment fault can be excluded from the PDFC of the sequential circuit. Again, the FSC values and the algorithm described in Section 5.2.3 are used to efficiently process combinational paths. If FSC values are used to exclude the segment fault from PDFC_S, then the segment fault is an untestable segment fault.

(3) If the segment starts at a flip-flop and terminates at a PO, the *Backward* phase is entered. Again, a finite number of frames in the iterative array model of the circuit is considered. The frames are labeled $t_{-k}, t_{-k+1}, \ldots, t_0$. The target segment is in frame t_0. All paths in the iterative array model that terminate at the target segment are implicitly examined. If none of these paths has to be included in the PDFC of the iterative array model, then the segment fault can be excluded from the PDFC of the sequential circuit. Again, if FSC values are used to exclude the segment fault, then the fault is classified as untestable.

(4) If the segment starts at a flip-flop and terminates as an input to a flip-flop, then the *Forward* phase is employed. If the segment fault cannot be excluded from the PDFC_S or the fault cannot be classified as untestable, then the *Backward* phase is employed.

The following example illustrates the mechanics of the above algorithm. To keep the discussion simple, the FSC values are not used.

EXAMPLE 5.20 Consider the circuit in Figure 5.18(a). Let the target segment fault be $\{\uparrow, chi\}$. The corresponding segment starts at the flip-flop c and terminates as an input to the flip-flop b. If FSC values are not used, then it can be shown that the segment fault is included in PDFC_C. Therefore, the *Forward* phase is entered. The iterative array model consisting of three frames labeled t_{-1}, t_0 and t_1 is shown in Figure 5.21. The *Forward* phase adds the frame t_1. The paths in frame t_1 that begin from the target segment are considered first. If segment $b_1e_1f_1$ in frame t_1 is considered, then a PO is reached. Implications of the mandatory assignments for the target segment fault are not sufficient to exclude segment $b_1e_1f_1$ from the PDFC of frame t_1. Therefore, based on the *Forward* phase, the target segment fault may have to be included in the PDFC_S. Next, the *Backward* phase is entered. Time-frame t_{-1} is added

PATH DELAY FAULT CLASSIFICATION 75

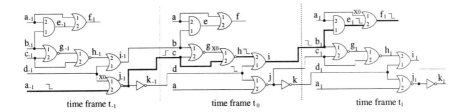

Figure 5.21. Untestable path delay faults.

and paths terminating at the target segment are examined. Frame t_{-1} has a segment $a_{-1}j_{-1}$ that starts at a PI and terminates at the target segment. Mandatory assignments in frame t_{-1} are again, not sufficient to exclude the segment $a_{-1}j_{-1}$ from the PDFC of frame t_{-1}. Therefore, the target segment fault is included in PDFC_S. Note that if FSC values are used, then the target segment fault can be classified as untestable. Also, a branch and bound algorithm can be used instead of implications to more accurately determine the status of the segment fault. However, this requires significant computing resources.

A partial path or segment-based strategy was also used by Agrawal *et al.* [3]. The sequential path test conditions were modeled by an equivalent single stuck-at fault and then a sequential circuit test generator was used.

5.5 SUMMARY

Unlike the stuck-at fault tests that can unconditionally detect the target faults, tests for delay faults differ in their levels of quality. This is because the detection of delay defects depends on the circuit timing, as well as on the fault model and the fault size. The path delay faults can be classified into several classes with respect to the test quality. The robust and validatable non-robust tests have the highest quality because they can detect a fault independent of the delays on the signals outside the target path. The non-robust tests can detect faults if certain other paths in the circuit are not faulty. The functional sensitizable paths may affect the performance only if there are other faults that simultaneously exist in the circuit.

Testing a path delay fault with the most stringent sensitization condition under which a test exists may lead to erroneous conclusions about

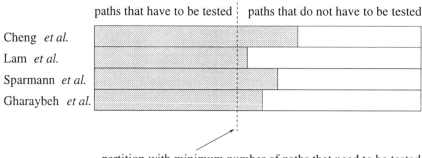

Figure 5.22. Comparison of different techniques described in Section 5.2.

the timing correctness of the design. The main problem is in the use of the oversimplified fault model that cannot take into account the dependence of the path delay on the applied test pattern. For example, Pierzynska and Pilarski [117] have shown that a path can have a considerably longer delay if it is tested non-robustly than if it is tested by a robust test. In deep submicron designs such effects can no longer be tolerated. The fault models have to be updated to better reflect the nature of defects.

Research on the path delay fault testing shows that there exist paths that can never independently impact the performance of the circuit and they do not have to be tested. Several different methods [32, 53, 92, 140] have been proposed for finding the set of faults that do not need to be tested and they differ in the size of the identified set as well as in their efficiency and ability to handle large designs. Figure 5.22 shows a comparison of the path delay fault classification methods described in Section 5.2.

The size of the primitive fault set does not depend on the specific technique that is used to identify the primitive faults. One way to test all primitive faults is to select one path from each primitive fault and show that it is not faulty. Selecting a more than minimal set of single paths to cover all primitive faults would result in some primitive faults being tested more than once.

6 DELAY FAULT SIMULATION

As with any other fault model, delay fault simulation requires less computational effort than delay fault test generation. Therefore, after a test is generated fault simulation should be performed to cover as many faults as possible with the same test. Delay fault simulation can also be performed with functional, random, stuck-at or any other available set of vectors to determine the delay fault coverage and reduce the delay test generation effort. This chapter discusses methodologies for simulating transition, gate, path and segment delay faults in combinational and sequential circuits.

The popularity of the transition fault model mainly comes from the availability of tools for testing and simulating stuck-at faults that can easily be adapted to test and simulate transition faults. On the other hand, simulating gate delay faults requires the consideration of delays and determination of threshold values of the faults detected by the given test. Transition and gate delay fault simulators deal with a linear number of faults in terms of the number of gates in the circuit. Path delay fault simulators have to handle a possibly exponential number of faults (in terms of the number of gates). In general, two kinds of path delay fault simulators have been developed: enumerative and non-

78   CHAPTER 6

enumerative. Enumerative fault simulators enumerate both the faults in the fault list and the detected faults, while the non-enumerative do not. Therefore, the non-enumerative simulators can handle a larger number of faults. This chapter describes several enumerative and non-enumerative path delay fault simulators. Some of the presented methodologies give a pessimistic estimate on the fault coverage, while others can compute the exact coverage. Segment delay faults are introduced as a trade-off between the transition and path delay faults. This chapter presents a methodology for simulating these faults.

6.1   TRANSITION FAULT SIMULATION

Fault simulation for transition faults can be performed using an adapted stuck-at fault simulator that can identify logic gates with transitions caused by an applied test pattern (relative to the preceding pattern). Simulating transition faults has been addressed in [30, 135, 151]. Waicukauski *et al.* [151] use an enhanced parallel-pattern, single fault propagation stuck-at fault simulator [150] to simulate transition faults in combinational circuits. The flowchart of the algorithm is shown in Figure 6.1 [151]. The fault list initially contains the slow-to-rise and slow-to-fall faults on each signal in the circuit. Next, the transition fault collapsing is performed. The rules for finding equivalence classes for transition faults are given in Section 3.1. The fault simulator combines the concepts of multiple pattern evaluation and single fault propagation. It simulates a set of patterns per pass. Single fault propagation minimizes the number of gate evaluations to determine if a given fault is detectable with the set of simulated patterns. Fault values are calculated starting from the fault location in an even-driven fashion. This algorithm can also be applied for simulating transition faults in sequential circuits under the slow-fast-slow testing strategy (see Chapter 2).

6.1.1   Simulating transition faults in sequential circuits

Fault simulation of transition faults in sequential circuits when the clock is applied at the circuit's rated speed has been addressed by Cheng [30]. The transition fault model for at-speed testing scheme has been discussed in Section 3.1. The inputs to the Transition Fault SIMulation algorithm (TFSIM) include: the gate-level circuit description, the input sequence and the sizes of transition faults for simulation. The overall flow of is shown in Figure 6.2. It is based on PROOFS stuck-at fault sim-

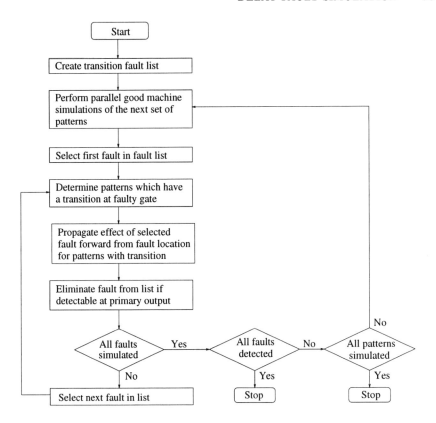

Figure 6.1. Parallel-pattern, single fault propagation transition fault simulation algorithm.

ulation algorithm for sequential circuits [111]. PROOFS combines the advantages of differential fault simulation, single fault propagation and parallel fault simulation. To fully utilize the bits of a computer word, faults are dynamically grouped to simulate several faulty machines at once (a group represents 32 faulty machines). The dynamic strategy avoids wasting space in the machine word for faults already detected. For each test vector, the algorithm performs the good circuit simulation and does the faulty evaluation for each fault group. TFSIM differs from PROOFS in the fault list generation, fault injection and storing of the states of faulty circuits.

Fault list generation. The fault list in TFSIM contains slow-to-rise and slow-to-fall faults on all signals in the circuit. However, since this transition fault model considers different fault sizes (specified as a num-

80   CHAPTER 6

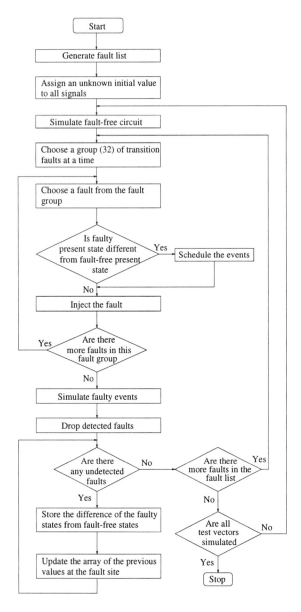

Figure 6.2.   Flowchart of the TFSIM algorithm.

ber of clock cycles), the sizes of transition faults that need to be simulated are passed on to the fault list generator. For example, if sizes 1 to 5 clock cycles are specified, the fault list generator will generate 5 slow-to-rise and 5 slow-to-fall faults for each signal. Fault collapsing is then performed among the faults of the same size.

FAULT INJECTION. After scheduling the state events that contain the differences of the present states between the fault-free and faulty circuits, fault injection is performed. There are two situations where a fault event should be injected: (1) A transition occurs at the fault site. The transition should be suppressed for the current time-frame. (2) A transition occurred at the fault site in the previous time-frame(s) and was delayed until the current time-frame. To examine these conditions, it is necessary to know the value(s) of the faulty signal in the previous time-frame(s). If the size of the target transition fault is k clock cycles, the values of the faulty signal in the previous k time-frames have to be known. These values are recorded and updated at the end of the simulation for each vector. At the beginning of the simulation for the first vector, the initial value of the faulty signal is assumed to be an unknown value (X).

A traditional implementation of fault injection may use a flag for each gate. The flag indicates whether or not the associated gate is faulty. This implementation requires that the flags be examined for every gate evaluation even though only one gate is faulty. The fault injection method proposed in [30] does not require the use of special flags. For each fault, an extra gate is inserted into the circuit. The concept of inserting an extra gate for fault injection of stuck-at faults was used in PROOFS [111]. To inject a slow-to-rise fault of size k clock cycles, a $(k+1)$-input AND gate is inserted at the faulty signal. This is illustrated in Figure 6.3. The values of the first k inputs of the extra gate are set to the values of the faulty signal in the previous k time-frames. Note that if the value of signal A is a logic 0 in any of the previous k time-frames, the value of signal B in the current time-frame will be a logic 0. If the value of signal A is a logic 1 in all previous k time-frames, the value of signal B in the current time-frame will be the current value of signal A. Similarly, to inject a slow-to-fall fault of size k clock cycles, a $(k+1)$-input OR gate is inserted at the faulty line and the values of first k inputs to the extra gate are set to the values of the faulty signal at the previous k time-frames.

EXAMPLE 6.1 Consider the example given in Figure 6.4. The values at signal B can be expressed in terms of the values at signal A as:

$$B_n = A_n \wedge A_{n-1} \wedge A_{n-2}$$

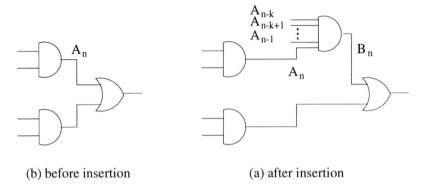

(b) before insertion   (a) after insertion

Figure 6.3. Fault injection at time-frame n for a slow-to-rise fault at signal A of size k clock cycles.

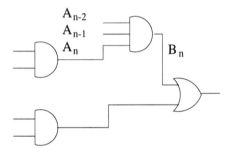

Figure 6.4. Fault injection example.

Therefore, the suggested fault injection technique produces a correct waveform at the fault site.

When simulating 32 faulty circuits in parallel, except for the injected fault, the values of the first k inputs to the extra gate are set to a logic value if an AND gate is inserted and to a logic value 0 if an OR gate is inserted.

Storing states of faulty circuits. At the end of the simulation of a vector, in addition to storing the difference between the faulty state and the fault-free state, the value at the faulty signal (the $(k+1)$-th fanin of the inserted gate) should also be stored. For each fault of size k clock cycles, an array of size k is used to store the values of the faulty signal in k previous time-frames. The arrays are updated for all undetected faults at the end of the simulation for each vector. These arrays are used for fault injection as described above.

The TFSIM algorithm has a relatively low memory requirement. In addition to the memory for storing the values of the fault-free circuit for all signals in one time-frame and the memory for storing the differences between the fault-free and each undetected faulty circuit at the next state signals (as used in PROOFS [111]), TFSIM requires an array of k bytes for each undetected fault of size k clock cycles to store the values of the faulty signal in the previous k time-frames.

In TFSIM, an unknown initial value (X) is assumed for each signal. Under this assumption, a transition fault may prevent a circuit from initialization. As an illustration, consider transition faults of size one clock cycle. In three-valued logic, there are nine possible combinations of the previous value and current value at a signal: 00, 01, 0X, 10, 11, 1X, X0, X1 and XX. If the value-pair at the fault site is 01, a rising transition occurs and the slow-to-rise fault should be injected. The current faulty value becomes 0. If the value pair is X1, 0X or XX, a rising transition may or may not occur, depending on the power-up states of the flip-flops. For example, if the faulty value-pair is X1 before fault injection, the slow-to-rise fault is injected and the current faulty value becomes X. Similarly, if the faulty value-pair is 0X before fault injection, the slow-to-rise fault is injected and the current faulty value becomes 0. If the value-pair of the faulty signal is X1 before injection, the current value of the signal becomes X after the slow-to-rise fault is injected. Similarly, if the value-pair of the faulty signal is X0, the current value of the signal becomes X after the slow-to-fall fault is injected. Table 6.1 summarizes the actions of fault injection for all value-pairs. The following example illustrates that a transition fault may prevent a circuit from being initialized, and thus it is untestable, while the corresponding stuck-at fault is initializable and testable.

EXAMPLE 6.2 Consider a three-bit counter with a reset input. The most significant bit is a primary output. The input sequence is the reset vector followed by a sequence of clock pulses. Table 6.2 lists the states of the fault-free circuit, the states of the circuit with a slow-to-fall fault of size one clock cycle at the least significant bit (denoted as LSB in the table), and the states of the circuits with stuck-at-1 and stuck-at-0 faults at the least significant bit. As it can be seen from the table, neither the stuck-at-0 nor the stuck-

Table 6.1. Fault injection.

Value-pair at the faulty signal		
before injection	after injection	
	slow-to-rise	slow-to-fall
00	-	-
01	00	-
0X	00	-
10	-	11
11	-	-
1X	-	11
X0	-	XX
X1	XX	-
XX	-	-

- No injection

at-1 fault prevents the circuit from initialization, while the circuit with the slow-to-fall fault is not initialized and the fault remains undetected.

Table 6.2. Faulty states for unknown initial values.

input event	fault-free circuit	LSB slow-to-fall	LSB stuck-at-1	LSB stuck-at-0
initial state	XXX	XXX	XXX	XXX
reset	000	00X	001	000
clock	001	0XX	011	000
clock	010	XXX	101*	000
clock	011	XXX	111	000
clock	100	XXX	001	000*
clock	101	...	...	...
clock	110	...	...	...
clock	111	...	...	...
clock	000	...	...	...

* Fault is detected

However, if we enumerate the two possible initial binary values (0 and 1) at the fault site and simulate them separately, the problem is solved for this example. Table 6.3 lists the faulty states for different initial values. The sequence detects the slow-to-fall fault in both cases at the same vector.

Table 6.3. Faulty states for different initial values.

input event	fault-free circuit	LSB slow-to-fall initial value = 0	LSB slow-to-fall initial value = 1
initial state	XXX	XX0	XX1
reset	000	000	001
clock	001	001	001
clock	010	011	011
clock	011	100*	101*
clock	100	101	110
clock	101	111	...
clock	110	...	...
clock	111	...	...
clock	000	...	...

* Fault is detected

As illustrated, the reported transition fault coverage may be too pessimistic if an unknown initial value is assumed for the faulty signal. A higher fault coverage can be achieved if each transition fault is treated as two faults: one that assumes an initial value 0 at the faulty signal and the other that assumes an initial value of 1.

6.2 GATE DELAY FAULT SIMULATION

For gate delay faults, the detection of a fault with a given test depends on the size of the fault. A test might detect faults of one size while faults of some smaller size stay undetected with the same test. Therefore, in addition to answering the question of whether or not the given fault is detected by the test, gate delay fault simulators have to also determine the sizes of the detected faults, i.e., the fault detection threshold.

EXAMPLE 6.3 Consider the circuit in Figure 6.5. Let the target gate delay fault be b slow-to-fall. Let the rising and falling delay of

86   CHAPTER 6

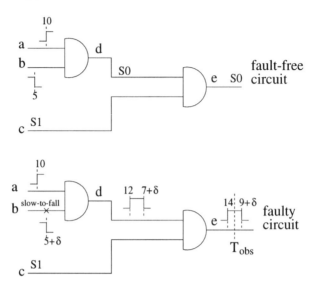

Figure 6.5. Gate delay fault simulation and determining the detection threshold.

both AND gates in the figure be 2 time units. Also, let the output be observed at time $T_{obs} = 15$. In the fault-free circuit, there will be a stable 0 value on signal d and a stable 0 value at signal e. If the falling transition at signal b is delayed for $\delta \geq 5$ there will be a static 0-1-0 hazard at signal d. If this spike is wide enough, it might propagate through the second AND gate. If we assume that any spike whose width is greater than or equal to 2 will propagate to signal e, then the test shown in the figure will detect any slow-to-fall fault at b with size $\delta \geq 6$.

Fault simulators for gate delay faults have been reported in [22, 43, 45, 71, 105, 126]. Carter et al. [22] propose a method to estimate the gate delay fault coverage under the condition that the delay fault sizes are known. Signal values are represented using "simplified signal waveforms" [22]. For each signal s, the simplified waveform is characterized by two numbers: **earliest arrival time** $EA(s)$ and **latest stabilization time** $LS(s)$. For each two-pattern test, $V = \langle v_1, v_2 \rangle$ the signal is assumed to have its **initial value** $IV(s)$ (value under vector v_1) for $t \leq EA(s)$ and its **final value** $FV(s)$ (value under vector v_2) for $t \geq LS(s)$. For $EA(s) \leq t \leq ST(s)$ the signal is assumed to have an unknown value. For example, a signal waveform shown in Figure 6.6(a) is simplified using the waveform given in Figure 6.6(b). Simulation of the fault-free circuit involves computing the EA and LS times for all

Figure 6.6. Representing signal values: (a) detailed and (b) simplified signal waveform.

signals in the circuit. This requires the knowledge about the rising and falling delays of gates in the circuit. Let

DAC (DAN) denote the shortest possible delay for the gate output to change away from controlling (non-controlling) value,

DSC (DSN) denote the longest possible delay for gate output to stabilize at controlling (non-controlling) value,

QCI (QNI) denote the set of all inputs to a given gate with controlling (non-controlling) initial values, and

QCF (QNF) denote the set of all inputs to a given gate with controlling (non-controlling) final values.

For a simple gate (AND, NAND, OR or NOR) the earliest arrival time and latest stabilization time can be computed as [22]:

$$PreEA(s) = \begin{cases} \max_{s \in QCI}(EA(s)) + DAC, & \text{if } QCI \neq \emptyset \\ \min_{s \in QNI}(EA(s)) + DAN, & \text{if } QCI = \emptyset \end{cases}$$

$$PreLS(s) = \begin{cases} \min_{s \in QCF}(LS(s)) + DSC, & \text{if } QCF \neq \emptyset \\ \max_{s \in QNF}(LS(s)) + DSN, & \text{if } QCF = \emptyset \end{cases}$$

If for signal s, the calculation of $PreEA(s)$ and $PreLS(s)$ shows $PreEA(s) \geq PreLS(s)$, then signal s has a stable value [22]. Therefore,

$$EA(s) = \begin{cases} PreEA(s), & \text{if } PreEA(s) < PreLS(s) \\ +\infty, & \text{if } PreEA(s) \geq PreLS(s) \end{cases}$$

88   CHAPTER 6

$$LS(s) = \begin{cases} PreLS(s), & \text{if } PreEA(s) < PreLS(s) \\ -\infty, & \text{if } PreEA(s) \geq PreLS(s) \end{cases}$$

EXAMPLE 6.4 Consider the AND gate in Figure 6.7. Let the delays of the AND gate be: $DAC = 1$, $DAN = 3$, $DSC = 6$ and $DSN = 4$. Let the signal waveforms on the inputs of the AND gate be as shown in the figure. Then, $QCI = \emptyset$ and $QCF \neq \emptyset$. The earliest

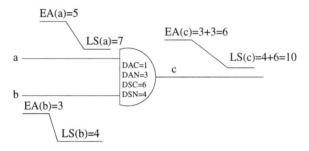

Figure 6.7. Calculating the EA and LS values at the output of an AND gate.

arrival time at the output c can be found as $EA(c) = \min\{5, 3\} + DAN = 3 + 3 = 6$, while the latest stabilization time is $LS(c) = \min\{4, 7\} + DSC = 4 + 6 = 10$.

Fault injection involves adding extra delay to the EA and LS times of the faulty signal. Next, the EA and LS times of the faulty and fault-free circuits at each signal need to be compared. If there is a difference, the fault is propagated further. Let T_{obs} denote the time when the primary outputs are observed after the application of the second vector at the primary inputs, and let $EA(z)$ and $LS(z)$ denote the earliest arrival and latest stabilization times at a primary output z, respectively. Then, if for a given fault under a given test there is a transition at z and $EA(z) \geq T_{obs}$, the fault is said to be *definitely detected* at primary output z. The fault is considered *possibly detected* at z if there is a transition at z and $EA(z) < T_{obs} < LS(z)$ [22].

Iyengar et al. [71] propose a method for computing a fault detection threshold for a given vector pair and a given fault. All faults larger than the detection threshold are considered detected by the test set.

6.3   PATH DELAY FAULT SIMULATION

Test generation for path delay faults is a complicated and computationally expensive task. Using path delay fault simulators can reduce the test generation efforts. There has been a lot of research effort directed towards developing path delay fault simulators [14, 23, 46, 54, 61, 68, 75, 77, 98, 121, 124, 139]. Depending on whether or not they require enumeration of the faults in the fault list and in the list of detected faults, path delay fault simulators can be grouped into enumerative and non-enumerative types. This section describes several enumerative and non-enumerative path delay fault simulators for combinational and sequential circuits.

6.3.1 Enumerative methods for estimating path delay fault coverage

Enumerative path delay fault simulators operate with a list of path delay faults and a list of detected faults. Usually, paths in the fault list are represented as a sequence of signals that belong to the path. Once a path is tested by a given test sequence it needs to be compared to the list of previously tested and stored path delay faults to find the fault coverage. This in turn, requires comparing sequences of signals. To reduce the memory requirements and increase the speed of the path delay fault simulation process, paths can be represented using *path numbers* [14, 123]. Faults are then represented as pairs (path number, transition type), where the transition type is rising or falling. By assigning a unique path number to each path in the fault list, paths tested by the given vector sequence can be found by comparing the path numbers rather than comparing a sequence of signal numbers.

Combinational circuits. Fault simulation for path delay faults in combinational circuits has been first addressed by Smith [139]. A six-valued algebra for simulating robust path delay faults is proposed. It consists of the following values: S0, S1, P0, P1, -0 and -1. Symbol S is used for signals with stable value during the application of a two-pattern test. These signals do not have any hazards. The output of a gate is assigned value P if the transition on it cannot occur before all inputs with value P have changed from their initial values. These signals have different initial and final values but hazards might exist before the signal stabilizes in its final value. Value − is used for signals that cannot be assigned either value S or value P. These signals can have none, one or more transitions. The six-valued logic system is illustrated in Figure 6.8(a). The implication tables for this system for AND, OR and NOT gates are given in Figure 6.8(b). For each two-pattern test,

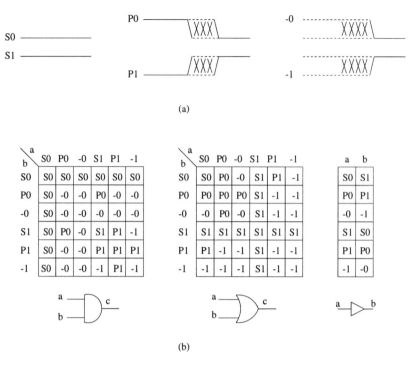

Figure 6.8. Implication tables for six-valued logic.

the simulation algorithm [139] traces all tested paths in the fault list and the paths with value P1 or P0 on all signals are marked as detected and removed from the fault list. The fault list contains all or selected paths in the circuit. Parallel processing of patterns is used to reduce the fault simulation time.

Kapoor [77] proposes an efficient enumerative method for computing the exact path delay fault coverage. In this method closed intervals consisting of pairs of integers represent sets of consecutive path delay faults. Each fault is assigned a unique path number [14] and path delay faults with numbers $i_1, i_1 + 1, \ldots, i_2$ are represented as interval $[i_1, i_2]$. A tree-based data structure is then used to store and manipulate the intervals during fault simulation.

Most path delay fault simulation methods for combinational circuits assume that after the application of the first vector, v_1 of each vector pair $V = \langle v_1, v_2 \rangle$, the circuit is allowed enough time for all signal values to stabilize before applying the second vector, v_2. In practice, tests are often applied at-speed, which means that a rated clock is applied after application of each vector in the vector pair V. Path delay fault simulation under at-speed vector application for robust faults in combinational circuits has been considered by Hsu and Gupta [68].

Sequential circuits. In sequential circuits, the source of a path can be a primary input or a flip-flop and the destination of a path can be a primary output or a flip-flop. Fault from the target path can affect some flip-flops other than the destination flip-flop or destination primary output. Therefore, the fault simulators for non-scan sequential circuits have to decide how to update the states of the flip-flops other than the destination flip-flop after the activation of the target fault. For example, if during simulation a rising transition arrives at the destination flip-flop, the correct value of the flip-flop will be 1. If the rising transition arrives late, the destination flip-flop will latch in a value 0. If a transition also reaches some flip-flops other than the destination flip-flop, these flip-flops might latch in an incorrect value [4, 24]. Also, in non-scan sequential circuits, the path delay fault simulation procedure depends on the assumed test application scheme (see Chapter 2). If the slow-fast-slow clock testing scheme is assumed, the fault is activated only once during the application of the test sequence for that fault. In the at-speed scheme, the fault can be activated many times during the application of the test sequence. Path delay fault simulators for sequential circuits have been proposed by several research groups [14, 23, 124].

Chakraborty et al. [23] address path delay fault simulation for both slow-fast-slow and at-speed testing strategies. Two different fault models are considered. In the first model, it is assumed that the fault effect reaches only the destination flip-flop. All other flip-flops are assumed to latch in their fault-free values. Tests generated for this fault model can be invalidated by the presence of some faults other than the target fault. In the second fault model, only those flip-flops that have stable values without hazards during fault activation are updated with their fault-free circuit values. All other flip-flops (except the destination flip-flop) are assigned an unknown value in the faulty circuit. Tests generated for this fault model cannot be invalidated. However, the fault coverage obtained with this fault model can be pessimistic.

At-speed path delay fault simulation has also been considered by Bose et al. [14]. This fault simulator is based on the six-valued logic system shown in Figure 6.8. Both robust and non-robust detection modes for path delay faults are simulated. For non-robust path delay fault propagation, the only difference in the implication table for AND gate is for the case ($a = $ P0, $b = $ -1) for which the output c evaluates to P0. This is because it is assumed that input b settles in its final value 1 quickly and the transition from input a propagates to the output c. Similarly, for the OR gate, case (a = P1, b = -0) evaluates to P1. For robust simulation, the *optimistic update rule* [14] is used to update the values of the flip-flops after the fault activation. According to this rule, all flip-flops with non-steady values when the fault is activated are updated with their fault-free circuit values given that they are destinations of at least one robustly activated path. The flip-flops with non-stable values resulting from a non-robust propagation are updated with unknown value. The flip-flops with steady values are assigned their fault-free values. It can be proven that the faults found detectable using the optimistic update rule are guaranteed to cause a failure and cannot be masked by other faults in the circuit [14]. The fault coverage derived using this rule is higher than that derived by assuming that all flip-flops with non-steady values be updated with an unknown value. For non-robust simulation, all flip-flops, other than the fault destination flip-flop, are updated with their fault-free values.

A variation of the slow-fast-slow clocking strategy for simulation of non-scan sequential circuits has been considered by Pomeranz and Reddy [124]. For a test sequence consisting of k vectors all possible $(k-1)$ schemes that have a single fast clock are simulated in parallel. Only $(k-1)$ schemes are considered since the first vector is needed for initializing the circuit. In each of the $(k-1)$ schemes the fast clock is applied for a different test vector. The idea is that once a test sequence is generated some additional faults can be detected by the same sequence if the fast clock is applied during a different clock cycle. This simulator can also be used for randomly generated test patterns for which the fast clock can be randomly assigned. All flip-flops other than the destination flip-flop are assigned an unknown value in the faulty circuit. Pomeranz and Reddy [124] also consider the application of multiple fast clocks.

6.3.2 Non-enumerative methods for estimating path delay fault coverage

The number of path delay faults can be exponential in terms of the number of signals in the circuit. Therefore, fault simulation methods that rely on enumerating path delay faults both in the fault list and in the list of detected faults usually cannot process large circuits.

EXAMPLE 6.5 Consider a circuit that contains n blocks of logic of type C shown in Figure 6.9 [121]. The total number of faults, the number of simulated tests and the number of detected faults for different values of n are shown in Table 6.4 [121].

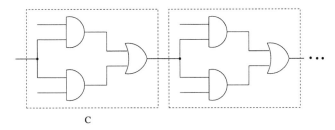

Figure 6.9. Example circuit with exponential number of faults.

Table 6.4. Fault simulation results for the circuit in Figure 6.9.

n	tests	total faults	detected faults
20	41	6 291 452	3 145 726
30	61	6 442 450 940	3 221 225 470
40	81	6 597 069 766 652	3 298 534 883 326
50	101	6 755 399 441 055 740	3 377 699 720 527 870

The problem can be alleviated by using non-enumerative methods for estimating the path delay fault coverage. Non-enumerative methods for combinational circuits have been proposed by several research groups [54, 61, 76, 121].

Pomeranz and Reddy [121] propose a method that can estimate the fault coverage in time which is polynomial in the number of signals in the circuit. The computed fault coverage is pessimistic in the sense that the exact fault coverage is never smaller than the estimated one. The quality of the estimation can be improved by increasing the polynomial complexity of the method.

Estimating the path delay fault coverage requires knowledge about the total number of faults and the number of faults detected by the given test set. The total number of path delay faults can be found in linear time in the number of signals in the circuit [121].

Counting the number of path delay faults. To find the number of paths, the gates are processed in a topological order from primary outputs towards primary inputs. During processing, all signals in the circuit are assigned integer values as follows: (1) Each primary output is assigned value 1. (2) Each input to a gate is assigned the same value as the gate's output. (3) The values on the fanout stems are obtained as a sum of the values on all fanout branches. The number of paths in the circuit can be found as a sum of the values assigned to the primary inputs and the number of path delay faults is twice the number of paths. The following example illustrates the path counting procedure.

EXAMPLE 6.6 Consider the circuit in Figure 6.10 [121]. The value assigned by the path counting procedure is shown within square brackets next to each signal. Primary outputs j and k are assigned value 1. All gate inputs are assigned the same value as the gate output. The value on the fanout stem f (h) is obtained as a sum of the values on the fanout branches f_1 (h_1) and f_2 (h_2). The sum of values assigned to the primary inputs $4 + 3 + 1 + 2 + 1 = 11$ is the number of paths in the circuit. Therefore, there are 22 path delay faults in this circuit.

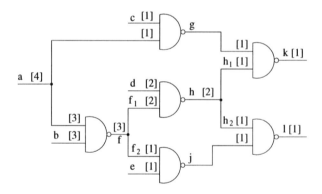

Figure 6.10. Counting the number of paths in a circuit.

Counting the number of detected faults for a given test. The number of path delay faults robustly detected by a two-pattern test, can be found without enumerating the faults by simulating the circuit and determining the signals that robustly propagate transitions [121]. After that, the path counting procedure can be adapted to count the robustly detected faults without enumerating them. The procedure for finding the signals that robustly propagate transitions under a given test is illustrated by the following example.

EXAMPLE 6.7 Consider the two-pattern test shown in Figure 6.11 [121]. The values assigned under the test are shown inside the small squares next to each signal. Only primary inputs b, c and e are assigned transitions under the test, while primary inputs a and d are assigned stable 1 (denoted S1) and stable 0 (denoted S0) values, respectively. The list of signals in parentheses indicates the inputs to the given gate that robustly propagate the transition to the gate output. Let a list associated with signal s be $R(s)$. The lists are assigned during simulation. Only those signals that have transitions can be assigned a non-empty list of signals. Each primary input that has a transition is assigned a list that contains the name of the primary input. Primary inputs that have stable values are assigned an empty list. For example, primary input a is assigned stable 1 value and $R(a) = \{\}$. Primary input b is assigned a falling transition and $R(b) = \{b\}$. Since the rising transition on signal f is due to the robustly propagated transition from input b, signal f is assigned list $R(f) = \{b\}$. Also, since the transitions on signals e and f both robustly propagate to signal i we get $R(i) = \{e, f\}$. The primary output k is assigned a transition that robustly propagates from signal i. Therefore, $R(k) = \{i\}$. This procedure ensures that for each signal s that is assigned a non-empty list, $R(s) \neq \{\}$, there exists a path from some primary input through which a transition robustly propagates to s.

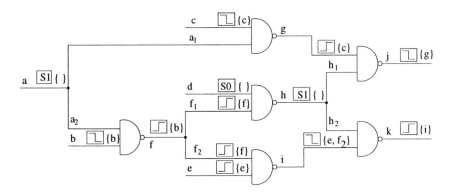

Figure 6.11. Tracing the paths detected by a two-input test.

This procedure can be extended to any other type of propagation conditions (non-robust, functional sensitizable, etc.).

The number of paths tested with a given test can be found using an adapted path counting procedure. This procedure assigns a non-zero value only to signals s for which $R(s) \neq \{\}$. All other signals are assigned value 0. The circuit is processed in topological order from the primary outputs towards the primary inputs and the signals with non-empty lists are assigned values as follows: (1) All primary outputs z with $R(z) \neq \{\}$ are assigned value 1. (2) An input m for which $R(m) \neq \{\}$ to a gate n is assigned the same value as signal n. (3) The value on a fanout stem is equal to the sum of the values assigned to its fanout branches. Then, the number of paths tested by the given two-pattern test can be found as the sum of the values assigned to the primary inputs.

EXAMPLE 6.8 Consider the circuit in Figure 6.12 [121]. The lists associated with signals are given in Figure 6.11. Signals with non-empty lists are shown with bold lines in the figure. All signals with

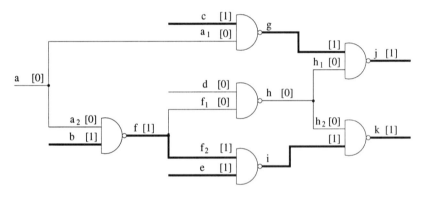

Figure 6.12. Counting the number of paths detected by a two-input test.

empty lists are assigned value 0. The signals with non-empty lists are assigned values in the same way as in the procedure for counting the number of paths in the circuit. There are three paths tested by the test shown in Figure 6.11: $\{\downarrow, cgj\}$, $\{\downarrow, bfik\}$ and $\{\uparrow, eik\}$.

Counting faults detected by a given test set. After each two-pattern test is simulated, the set of detected faults has to be compared to the previously detected faults to calculate the exact fault coverage. Only the new faults (faults not detected by some previously applied test) should be counted. However, the non-enumerative methods for estimating the fault coverage do not store detected path delay faults. To ensure that no detected path delay fault is counted more than once, Pomeranz and Reddy [121] require that every fault that is counted contains at least one pair of type (signal, transition) that was not included in any other previously tested fault.

EXAMPLE 6.9 Consider the two-pattern test in Figure 6.13 [121]. The values and the lists assigned to each signal under this test are shown in the figure. Let the test shown in Figure 6.11 be denoted as T_1 and the test in Figure 6.13 as T_2. Also, let test T_1 be the only test applied to the circuit before test T_2 is applied. Both tests detect path delay fault $\{\uparrow, eik\}$. To ensure that this path is not counted twice, only those paths tested by T_2 that cover at least one (signal, transition) pair not covered by T_1 are counted as detected by T_2. Test T_1 detects path delay faults that cover the following (signal, transition) pairs: $(b, \downarrow)$, $(c, \downarrow)$, $(e, \uparrow)$, $(f, \uparrow)$, $(f_2, \uparrow)$, $(g, \uparrow)$, $(i, \downarrow)$, $(j, \downarrow)$ and $(k, \uparrow)$. The only new (signal, transition) pairs covered by paths tested by T_2 are: $(a, \downarrow)$, $(a_1, \downarrow)$ and $(a_2, \downarrow)$. Therefore, any path counted as detected by test T_2 needs to cover at least one of these (signal, transition) pairs. Only path $\{\downarrow, afik\}$ satisfies this condition.

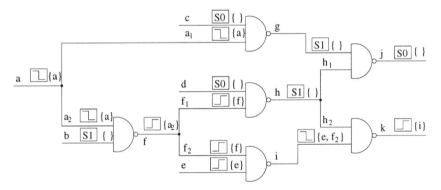

Figure 6.13. Finding the newly detected path delay faults.

The procedure for counting the total number of faults in the circuit can be further modified to count the detected faults covering at least one new (signal, transition) pair for each two-pattern test [121].

It is possible that some new path delay faults are detected by a test even if all (signal, transition) pairs that it covers have been previously covered. Therefore, the true fault coverage can be higher than estimated. The fault coverage estimate can be improved by partitioning the circuit into subcircuits such that each subcircuit has a unique set of paths [121]. Then, the new (signal, transition) pairs can be separately considered with respect to each subcircuit. To ensure that each subcircuit covers a unique set of paths, one may find a cut through the circuit such that every path in the circuit passes through exactly one signal in the cut. Each signal in the cut defines a subcircuit. Since some signals can belong to more than one subcircuit and the subcircuits cover a disjoint set of paths, the (signal, transition) pairs can be covered by more than one detected path given that these pairs appear in different subcircuits.

EXAMPLE 6.10 Consider the circuit in Figure 6.13 and cut $C = \{a_1, a_2, b, c, d, e\}$. This cut contains all primary inputs or their fanout branches and therefore, every path in this circuit passes though exactly one signal in C. Each signal in the cut defines a subcircuit (signals that are in the fanin or fanout cone of the given signal in the cut). The subcircuits for signals in C are the following:

a_1: $\{a, a_1, g, j\}$, c: $\{c, g, j\}$,
a_2: $\{a, a_2, f, f_1, f_2, h, h_1, h_2, i, j, k\}$, d: $\{d, h, h_1, h_2, j, k\}$,
b: $\{b, f, f_1, f_2, h, h_1, h_2, i, j, k\}$, e: $\{e, i, k\}$.

Since, for example, signal g is contained in two subcircuits each of the pairs $(g, \downarrow)$ and $(g, \uparrow)$ can be be covered by two different tested paths that can be counted for the fault coverage.

Finding a cut of maximum size is advantageous since it would result in a larger number of subcircuits. Kagaris et al. [76] propose a polynomial-time algorithm for finding a maximum cut of a combinational circuit such that every path in the circuit passes through exactly one signal in the cut. The accuracy of the fault coverage estimate can be further improved if more than one cut is considered [76, 121].

Another way to obtain a better lower bound on the fault coverage is to consider pairs of type (subpath, transition), where a subpath represents several consecutive signals and the given transition is applied at the source of this subpath [62, 63].

Non-enumerative exact methods for computing the path delay fault coverage through path counting have been proposed by Gharaybeh et al. [54, 56] and by Parodi et al. [114].

6.4 SEGMENT DELAY FAULT SIMULATION

The segment delay fault model has been proposed as a trade-off between the transition and path delay faults [64]. Heragu et al. [65] propose a fault simulator for segment delay faults. The segment length L is passed on as an input to the simulator. Therefore, the fault list for a chosen L contains all segment delay faults of length L and all path delay faults whose length is less than L. The simulator is based on a segment counting procedure [64], which reports the number of faults of length L, and on the segment labeling technique [65] that assigns a unique label to each detected segment fault.

As discussed in Chapter 5, many circuits have a large number of robustly and non-robustly untestable path delay faults that can affect the timing of the circuit (called functional sensitizable faults). To ensure that defects on these paths do not degrade the performance, multiple path delay faults have to be identified and tested. This might not be feasible for large designs. On the other hand, many functional sensitizable paths contain segments that can be robustly tested and it might be beneficial to test them [65].

EXAMPLE 6.11 Consider the circuit in Figure 6.14. Path delay fault $\{\downarrow, bceghi\}$ is neither robustly nor non-robustly testable but

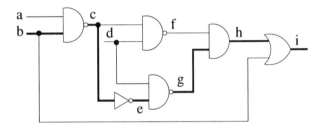

Figure 6.14. Functional sensitizable path delay fault consisting of robustly testable segment faults.

can affect the performance of the circuit if fault $\{\downarrow, dghi\}$ also occurs (see Section 5.3). However, segment delay faults $\{\downarrow, bce\}$ and $\{\downarrow, ghi\}$ are both robustly testable.

6.5 SUMMARY

Each fault model represents only an approximation of the physical defects. It is used to generate tests for detecting circuit malfunctions. Fault simulation is used to evaluate the generated tests, derive fault dictionaries or analyze the operation of the circuit under faulty conditions [1]. Delay fault simulation techniques have been developed for all popular delay fault models. These models and their associated test generation and fault simulation techniques deal with gate level representations of the circuit. Delay faults in designs that contain non-Boolean components like precharge or dynamic logic and bidirectional devices cannot be handled at the gate level [17]. Also, using gate level representations can lead to erroneous fault simulation results. Consider the example of

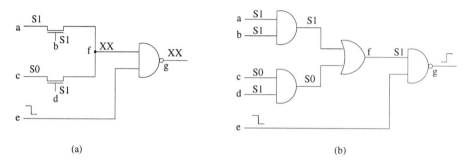

Figure 6.15. Example of erroneous path sensitization at the gate level.

an NMOS circuit and the two-pattern test shown in Figure 6.15(a) [17]. A gate level representation of this circuit is shown in Figure 6.15(b). Let the target fault be $\{\downarrow, eg\}$. The value of node f in Figure 6.15(a) is unknown since there is a short circuit between inputs a and c. Therefore, the target fault is not sensitized by the given two-pattern sequence. However, the same two-pattern sequence sensitizes the target fault in the gate level representation of the circuit as shown in Figure 6.15(b). These problems can be overcome if a switch-level representation of the circuit is used. Fault simulation of delay faults at the switch level has been addressed by Bose et al. [17].

7 TEST GENERATION FOR PATH DELAY FAULTS

This chapter concentrates on multi-value systems and methods proposed for generating tests for single and multiple path delay faults. The robust, non-robust, validatable non-robust and functional sensitizable faults are considered as single path delay faults. These paths usually can be tested with many different tests, i.e., there are many different robust tests for a robust testable path, many different non-robust tests for a non-robust testable path, etc. Since robust tests are guaranteed to detect a faulty target path independent of whether or not there are delay faults on paths other than the target path, most test generators do not differentiate between robust tests for a given path. These test generators do not use any timing information. On the other hand, some non-robust tests have a higher probability of detecting a faulty non-robust testable path than other tests. Similar argument holds for the functional sensitizable tests. The higher quality non-robust and functional sensitizable tests can be found by including the timing information into the test generation process. This chapter presents test generation algorithms that can produce high quality tests based on using the timing information for the non-robust and functional sensitizable faults. A non-robust test for a given target path becomes invalid if certain other paths in the circuit are defective. If the faults that may invalidate the non-robust test for the target path can be robustly tested, those robust tests along with the non-robust test for the target path form a validatable non-robust test (VNR). An algorithm for automatic generation of validatable non-robust tests is outlined in this chapter.

Testing only single path delay faults is not sufficient to guarantee the circuit performance. Some multiple delay faults (multiple primitive faults) can also affect the performance. Identifying and testing such faults for large multi-level designs is a hard task. An algorithm that can identify and test double primitive faults is presented.

7.1 ROBUST TESTS

Process variations usually affect delays of more than one gate or interconnect in the circuit. Therefore, it would be ideal if all path delay faults could be tested under conditions that are independent of the delays on signals and gates outside the target path. However, this is possible only for a subset of path delay faults called robust testable faults. The definition of the robust testable path delay faults has been given in Section 5.1.

Robust testable path delay faults were first considered by Smith [139]. A six-valued logic system was suggested to find the path delay faults tested by a given two-vector pattern. Test generation for robust path delay faults was addressed by Lin and Reddy [98]. A five-valued logic is proposed for generating tests for the robust testable paths. The logic system consists of values {S0, S1, U0, U1, XX}. These symbols represent the initial and final values of a signal under a vector pair $V = \langle v_1, v_2 \rangle$. Symbol S0 (S1) is used for signals for which the initial and final values are 0 (1). Signals with S0 or S1 values are assumed to be hazard-free under vector pair V. Symbol U0 (U1) represents signals for which the final value is 0 (1). The initial value can be 0 or 1 or there could be hazards on the signal before it settles to its final value. Symbol XX represents signals for which the initial and final values are unspecified. Figure 7.1 [98] illustrates the covering relations in the proposed logic system. Value U0 (U1) covers S0 (S1), while XX covers both U0 and U1. In this value system a transition can only be represented using U0 (for falling transition) and U1 (for rising transition). Lin and Reddy [98] use the convention that for the signals on the target path symbols U0 and U1 can represent only a falling or rising transition, respectively, while for the signals outside the target path these symbols are interpreted as

TEST GENERATION FOR PATH DELAY FAULTS   103

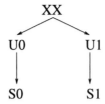

Figure 7.1.   Covering relations in the five-valued logic system.

defined earlier. The implication tables for AND, OR and NOT logic gates for the five-valued logic system are shown in Figure 7.2.

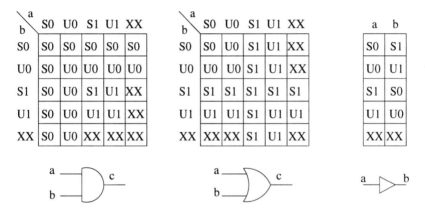

Figure 7.2.   Implication tables for AND, OR and NOT gates.

A two-vector test, $V = \langle v_1, v_2 \rangle$, is a robust test for a given path delay fault if it launches the desired transition at the source of the path and the path's off-inputs have logic values that are covered by the values given in Figure 7.3 [98]. For example, if the on-input to an AND or NAND gate is assigned a rising transition, the off-input can be assigned any value covered by U1, i.e., a value S1 or U1. As it can be seen from

gate type / on-input transition	AND NAND	OR NOR
Rising	U1	S0
Falling	S1	U0

Figure 7.3.   Values covering the off-input values.

the figure, in the proposed five-valued logic system the values on the off-inputs are uniquely determined. They satisfy the conditions given by Definition 5.1 for the robust off-inputs.

A PODEM-type test generation algorithm based on this logic system has been proposed by Patil and Reddy [116]. It starts with the list of paths for which tests have to be generated. Next, given the path and a transition at the source of the path, a list of signal values required for the on-inputs and off-inputs (according to the table in Figure 7.3) are generated. The test generator then attempts to justify the signals in the list using a PODEM-type algorithm [57]. After a partially specified test is generated, a heuristic procedure is used to convert the unspecified (XX) or partially specified (U0 or U1) values at primary inputs into specific values such that the probability of detecting the remaining path delay faults with the same test is maximized.

Test generation algorithms for robust path delay faults have also been proposed by many other researchers. For example, Fuchs *et al.* [50] propose a ten-valued logic system and a stepwise path sensitization procedure to generate robust tests. This procedure is capable of handling a large fraction of path delay faults. The experimental results on many benchmark and industrial designs show that usually only a small number of path delay faults can be tested by a robust test. Saldanha *et al.* [131] propose a robust test generation algorithm that employs a single stuck-at fault test generator on a modified network.

Recently, it was shown that a two-vector test for a robust path delay fault might not always excite the worse case delay of the target path [49, 117]. The off-input transitions and pre-initialization are shown to have significant impact on the delay of the target path. Chen *et al.* [28] propose a robust path delay fault test generator that incorporates these two additional considerations. It generates three-vector robust tests that excite the worst case delay of the target path.

7.2 HIGH QUALITY NON-ROBUST TESTS

Non-robust testable paths have been defined in Section 5.1. As was shown, a non-robust test becomes invalid if the transition on *any* non-robust off-input arrives later than the transition on the corresponding on-input. A non-robust testable path can usually be tested with many different non-robust (NR) tests. Test generation techniques for non-robust testable path delay faults have been proposed in the literature [50, 129]. However, these techniques cannot make a distinction between different non-robust tests for a given path. Cheng et al. [34] show that some non-robust tests have a higher quality than others. Their algorithm for generating high quality non-robust tests is based on including timing information into the test derivation process. A metric, called *robustness* is introduced to measure how close a given non-robust (NR) test is to being a robust test. Then, for each non-robust testable fault, this metric is used to generate a non-robust test with high robustness.

The following notation and definitions will be used to describe the algorithm for generating high quality non-robust tests [34]. Let $V = \langle v_1, v_2 \rangle$ be an input vector pair applied for delay testing of a given target path and let v_2 be applied at time $t = 0$. At some time after $t = 0$ the logic values on the signals in the circuit will become stable. It is assumed that the signal delays in the fault-free circuit are equal to the nominal signal delays.

DEFINITION 7.1 The time when the logic value on signal f becomes stable under v_2 is called the **arrival time** of f under v_2. The arrival time at signal f under v_2 in the fault-free circuit is denoted as $AT(f, v_2)$.

DEFINITION 7.2 For a given off-input g and its corresponding on-input f, the difference $AT(f, v_2) - AT(g, v_2)$ is called the **slack of the off-input g**.

A non-robust testable path under any test vector pair has at least one non-robust off-input while the rest are robust off-inputs (see Section 5.1.3). The number of non-robust off-inputs for a given target path can be different for different non-robust tests. Since the goal of the algorithm is to generate non-robust tests that are *more robust*, its first objective is to find a test with a minimal number of non-robust off-inputs for the given target path. Also, two non-robust tests with the same number of non-robust off-inputs can have different quality with respect to their effectiveness in detecting defects. Slack of the non-robust test is defined to help guide the test generation process towards generating higher quality non-robust tests.

DEFINITION 7.3 Let $g_1, g_2, \ldots, g_n$ denote the non-robust off-inputs for a given target path under a given vector pair V. Let $s_1, s_2, \ldots, s_n$ denote the slacks of those non-robust off-inputs. Then, the **slack of the non-robust test V** is defined as $\min_{i=1,\ldots,n} \{s_i\}$.

The second objective of the algorithm is to find the non-robust test with the maximum slack among all non-robust tests for a given path. If the slack at the non-robust off-inputs is larger, the probability that the non-robust off-input transition masks the on-input transition (due to the late arrival of transitions at the off-inputs) is lower. In other words, a non-robust test with a larger slack can tolerate larger variations of the arrival times at the non-robust off-inputs. For delay defects caused by process variations, the slack of a non-robust test should be closely related to the probability of fault masking at the non-robust off-inputs.

EXAMPLE 7.1 Consider the circuit in Figure 7.4(a). The path de-

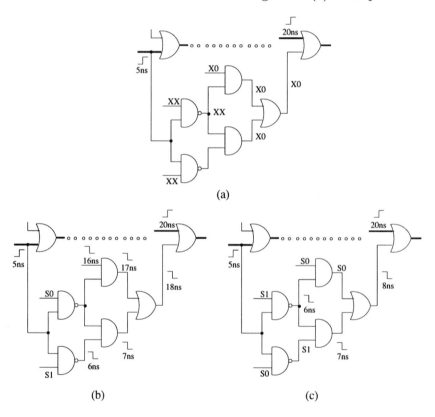

Figure 7.4. Different non-robust tests can tolerate different timing variations.

lay fault shown in bold is robustly untestable but is non-robustly testable. Figures 7.4(b) and 7.4(c) show two different non-robust tests for the same path. The test given in Figure 7.4(c) can tolerate a 12 ns delay variation on the non-robust off-input, while the test in Figure 7.4(b) can tolerate only a 2 ns delay variation.

Clearly, the test in Figure 7.4(c) has a higher quality than the test in Figure 7.4(b).

The **robustness** of a non-robust test is defined as its slack. The higher the robustness, the lower the probability of the test being invalidated. The robustness ranges from $-\infty$ to ∞. The robustness of a robust test is defined as ∞, while robustness of a vector pair which is neither a robust nor a non-robust test is defined as $-\infty$.

For each non-robustly testable path repeat:

1. Assign desired transitions at on-inputs.

2. Convert NR candidate off-inputs into robust off-inputs (if possible), one at a time. Leave the values of the candidate off-inputs which could not be converted into robust off-inputs unassigned. Timing information is used to determine the order of processing the candidates.

3. Convert the rest of the unassigned off-inputs into NR off-inputs one at a time and do backward justification. Timing information is used to determine the order of processing of the candidates and to make decisions during justification process.

4. Assign values to the remaining primary inputs that still have their values unassigned.

Figure 7.5. Summary of the algorithm for generating NR tests with high robustness.

7.2.1 Algorithm for generating non-robust tests with high robustness

The algorithm for generating non-robust tests with high robustness [34] consists of several steps. A brief summary of the algorithm is given in Figure 7.5. The detailed algorithm and a step-by-step example follow.

STEP 1: *Assign the on- and off-inputs to non-robustly sensitize the target path. Compute the earliest arrival time for each signal.*

To non-robustly test a path delay fault, all off-inputs must have a non-controlling value under vector v_2 and a transition must be created at the source of the path under test. These requirements have to be satisfied by any test vector and they are called **mandatory assignments** [81]. Hence, in the first step, all mandatory assignments and their implications [90, 91] are found.

Next, the earliest arrival time of each signal *restricted to these mandatory assignments* are computed. Note that certain input vector pairs may produce values that violate the given set of mandatory assignments (SMA) at some signals. The **earliest arrival time at signal f under a given SMA** is defined as the earliest arrival time at f among all vector pairs not violating the SMA.

EXAMPLE 7.2 Consider the circuit in Figure 7.6(a). The target path is shown in bold. The arrival times and the transi-

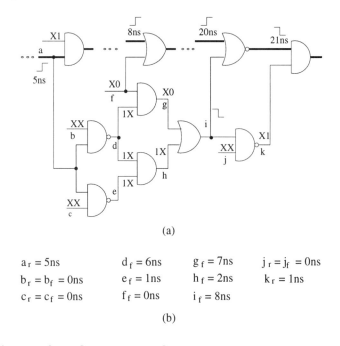

Figure 7.6. Computing the SMA and the earliest arrival times.

tions for the signals on the target path are as shown. Signals b, c, f and j are primary inputs and the transitions on them are assumed to arrive at time $t = 0$. All gates are assumed to have a unit delay for both, rising and falling transitions.

TEST GENERATION FOR PATH DELAY FAULTS   109

After assigning non-controlling values under the second vector, v_2, to all off-inputs and after implying these values, the obtained set of mandatory assignments is as shown. Next, the earliest arrival times for all signals under the given SMA are computed. These values are shown in Figure 7.6(b). Symbols s_f and s_r denote the arrival times of the falling and rising transitions on signal s, respectively. For example, $d_f = 6$ns means that the earliest arrival time for a falling transition on signal d under the given SMA is 6ns.

DEFINITION 7.4 *For a non-robust testable path, the off-inputs whose corresponding on-input has a ncv→cv value are called* **NR candidate off-inputs**.

EXAMPLE 7.3 In Figure 7.6(a), the off-inputs f, i and k are NR candidate off-inputs.

STEP 2: *Identify all NR candidate off-inputs for the target path and try to convert them, one at a time, into robust off-inputs by assigning them stable non-controlling values.*

To order the NR candidate off-inputs for processing, the slack of each NR candidate off-input (i.e., the difference between the arrival time of the corresponding on-input and the earliest arrival time of the transition on the NR candidate off-input) is computed. The NR candidate with the smallest slack is processed first. This is because a non-robust off-input with a smaller slack has a higher probability of masking its on-input transition than a non-robust off-input with a larger value of slack.

EXAMPLE 7.4 Consider again the circuit in Figure 7.6(a). The slacks of NR candidate off-inputs are: $slack(f) = 8$ns, $slack(i) = 12$ns and $slack(k) = 20$ns. Since the NR candidate off-input f has the smallest slack, signal f is first tried for conversion into a robust off-input by assigning a stable 0 (S0) value to it. The assignment of S0 value to f and its implications do not cause any conflicts and, therefore, signal f is successfully converted into a robust off-input.

Next, the earliest arrival times of signals are incrementally updated. This is needed because for certain signals the earliest arrival times may change due to the augmented set of mandatory assignments.

110 CHAPTER 7

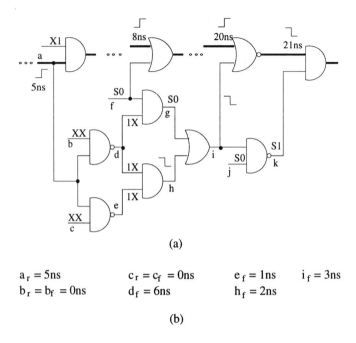

(a)

a_r = 5ns $c_r = c_f$ = 0ns e_f = 1ns i_f = 3ns
$b_r = b_f$ = 0ns d_f = 6ns h_f = 2ns

(b)

Figure 7.7. Converting non-robust off-inputs into robust off-inputs.

EXAMPLE 7.5 In the circuit in Figure 7.6(a), the earliest arrival time of signal g is removed from the set since g has a stable value under the new SMA and the earliest arrival time of signal i becomes $i_f = $ 3ns. The slacks of the NR candidate off-inputs now are: $slack(i) = $ 17ns and $slack(k) = $ 20ns. Therefore, the off-input i is selected as the next signal for processing. However, under the given SMA it is not possible to assign a stable non-controlling value to signal i and signal k is processed next. Off-input k can be converted into a robust off-input by assigning a stable 0 value to primary input j. The new SMA and the signal earliest arrival times are shown in Figure 7.7.

STEP 3: *Assign non-robust transitions to NR candidate off-inputs that cannot be converted into robust off-inputs under the current, partially assigned test T.*

To minimize the probability that the on-input transition is masked by the transition at a non-robust off-input, an attempt is made to find a test for which the arrival time of the transition at the off-input is the earliest possible. This can be achieved by using the calculated earliest arrival times to guide the justification process. To justify a transition at the output of a gate, among all inputs that can have a transition under current partial test, a transition is assigned to the input with the earliest arrival time. All other inputs to the gate are assigned stable non-controlling values. This backward justification process continues until either primary inputs are reached or a conflict occurs. In the latter case, the algorithm backtracks to the last decision point and justifies the transition at the input with the next earliest arrival time. The justification and backtracking process are very similar to those used in the stuck-at test generation algorithms.

In this test generation process, the values at internal signals and primary inputs are gradually assigned. Therefore, those NR candidate off-inputs that are processed later will have a smaller search space than those that were processed first. Thus, the most critical NR candidate off-inputs (ones with smaller slacks) are processed first.

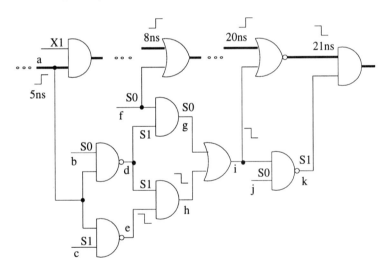

Figure 7.8. Backward justification and assigning unspecified values at PIs.

EXAMPLE 7.6 Consider the circuit in Figure 7.7(a). Signal i is a NR off-input for the target path (bold lines). The objective is to make sure that the falling transition on signal i arrives as early as possible. Since the value at signal h is justified using the calculated earliest arrival times for signals d and e, it can be seen that the transition on signal i will be the earliest if a

112   CHAPTER 7

falling transition is assigned to signal *e* and a stable 1 to signal *d*. This is illustrated in Figure 7.8.

STEP 4: *Assign the unspecified values at primary inputs.*

To obtain a completely specified test and to minimize the number of sensitized paths associated with the non-robust off-inputs, it is necessary to check if there are some primary inputs that do not have values assigned and they are assigned such that the number of transitions at primary inputs is minimized, i.e., value X1 is specified as 11, X0 is specified as 00, and XX as 00 or 11. Test compaction is not considered in this work. If test compaction is desired, this step of the algorithm can be omitted.

EXAMPLE 7.7 The only primary input that does not have completely specified values is the input *c* and in this step it is assigned a stable 1 value. The final values on all signals are shown in Fig. 7.8.

7.3 VALIDATABLE NON-ROBUST TESTS

The validatable non-robust path delay faults have been defined in Section 5.1.4. The validatable non-robust tests guarantee to detect a fault and, therefore, should be used for non-robustly testable paths whenever they exist.

There may exist many validatable non-robust tests for a given fault. To reduce the test set size it is of interest to find a test such that the number of paths that have to be robustly tested to validate the non-robust test for the target path is minimal. An algorithm for automatic generation of such a set of two-pattern tests is proposed in [34]. It consists of several steps.

STEP 1: *For each non-robust testable fault, convert the NR candidate off-inputs into robust off-inputs and derive a non-robust test with a minimal number of non-robust off-inputs.*

This is done using the procedure described in Section 7.2. After processing all NR candidate off-inputs, a partially specified non-robust test is obtained.

STEP 2: *Obtain a non-robust test T by specifying the unspecified values at the primary inputs such that the number of transitions at the primary inputs is minimized. Perform implications of T.*

STEP 3: *Examine the non-robust off-inputs and identify the paths that need to be robustly tested to validate T (see Figure 5.7).*

In developing the test T, the number of transitions at primary inputs is minimized. Therefore, typically only a very small number of partial paths that end at a non-robust off-input is sensitized. Thus, in Step 3 only a small number of paths need to be examined. This reduces the computational complexity and also reduces the cardinality of the validatable non-robust test.

There is a possible extension of this algorithm for identifying more VNR testable paths. The condition for robustly testing the sensitized partial paths (from primary inputs to non-robust off-inputs), can be relaxed so that they are tested under the VNR condition. However, if this extension is adopted, the following situation may occur: in generating a VNR test for path A, the VNR testability of path B is required and in generating a VNR test for path B, the VNR testability of path A is required. In this case, neither path is VNR testable.

Test generation for VNR path delay faults in non-scan sequential circuits has been addressed by Srinivas *et al.* [141]. VNR tests are dynamically generated using Boolean flags during the generation of robust tests.

As Cheng *et al.* [34] show, including the timing information into the process of non-robust test generation can substantially improve the quality of a non-robust test. However, experimental results on a large set of benchmark and industrial designs show that typically only a small number of paths in the circuit can be tested under non-robust and validatable non-robust criteria [50]. On the other hand, most circuits have a large number of functional sensitizable faults. These faults can, under certain conditions, affect the performance of the design. Therefore, to further improve of the quality of delay tests, it is necessary to derive tests for functional sensitizable paths.

7.4 HIGH QUALITY FUNCTIONAL SENSITIZABLE TESTS

The functional sensitizable path delay faults have been defined in Section 5.1.5. To *guarantee* that a given functional sensitizable (FS) path delay fault will not affect the performance of the circuit, a set of functional sensitizable tests must be applied. These tests will be considered in Section 7.5. In this section the goal is to generate a small number of tests (e.g., one) for a given FS path that is *most likely* to detect the fault. Different functional sensitizable tests have a different probability of detecting a defect on an FS path.

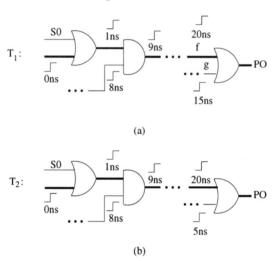

Figure 7.9. Different quality functional sensitizable tests.

EXAMPLE 7.8 Consider the functional sensitizable test T_1 shown in Figure 7.9(a). The target path is shown in bold and signal g is a functional sensitizable off-input under test T_1. Under this test, in the fault-free circuit, the rising transition at off-input g arrives at 15ns while the rising transition at the on-input f arrives at 20ns. The arrival time of the transition at the primary output will be determined by the earlier signal among the signals arriving at f and g. This means that, if in a faulty circuit, the transition at the off-input g arrives more than 5ns late, the fault on the target path will propagate to the primary output. Otherwise, the fault on the target path will not be detectable with test T_1. Next, consider the functional sensitizable test T_2 for the same target path (shown in Figure 7.9(b)). Signal g is again a functional sensitizable off-input. However, under test T_2 in the fault-free circuit, the transition at signal g arrives at 5ns. Therefore, in the faulty circuit, the rising transition on the off-input g needs to be more than 15ns late to observe the fault on the target path.

TEST GENERATION FOR PATH DELAY FAULTS   115

If it is possible to select only a small number of input vector pairs to test a functional sensitizable path, then one should select those tests that have a higher probability of making the defects on the FS path observable. In the above example, test T_1 is better than test T_2 since it requires smaller delay faults on signals outside the target path to detect a defect on the target path. Therefore, for a good FS test, the nominal arrival times of the FS off-inputs should be as late as possible, i.e., the slack of the FS off-input should be as small as possible. An algorithm for generating high quality functional tests will be reviewed in this section. The algorithm [34] uses the circuit timing information to derive high quality FS tests. For each tested FS path only one FS test is derived.

A necessary condition to detect a defective functional sensitizable path by a given test is that the transitions on the FS off-inputs arrive later than the transitions on the on-inputs. Hence, in general, if the number of FS off-inputs is larger, the probability of detecting a defect is smaller. The number of functional sensitizable off-inputs in a given path depends on the applied test vector pair. Therefore, one of the goals of the algorithm [34] is to minimize the number of FS off-inputs. Reduction of the number of FS off-inputs can be achieved by maximizing the number of off-inputs that are robust or non-robust. However, two FS tests with the same number of FS off-inputs can still have a different quality (as the above example shows).

DEFINITION 7.5 Let $g_1, g_2, \ldots, g_n$ denote the FS off-inputs under a given vector pair V, and let $s_1, s_2, \ldots, s_n$ denote the slacks of those FS off-inputs. Then the **slack of the FS test** V is defined as $\max_{i=1,\ldots,n} \{s_i\}$.

The second goal of the test generation algorithm is to find tests whose value of the slack is minimal. The slack in this case can be positive or negative. The algorithm searches for a test with the most negative slack, if it exists. Otherwise, it looks for one with the least positive slack. Such tests can tolerate only small timing variations on their FS off-inputs and are more likely to lead to the detection of a faulty FS path.

7.4.1 Algorithm for generating high quality functional sensitizable tests

The set of FS paths that need to be tested can be identified using any of the algorithms described in Section 3.2. In the process of generating non-robust tests the off-inputs whose corresponding on-input has a

116   CHAPTER 7

ncv→cv transition were called NR candidate off-inputs. In the context of generating tests for FS paths, such off-inputs are redefined as follows:

DEFINITION 7.6 *The off-inputs whose corresponding on-input has a ncv→cv transition are called* **FS candidate off-inputs**.

The algorithm for generating high quality functional sensitizable tests for a given FS path consists of several steps [34].

STEP 1: *Find the FS candidate off-inputs.*

STEP 2: *Try converting FS off-inputs into robust off-inputs one at a time.*

Depending on the transitions assigned under the given test vector, an FS candidate off-input can become either robust, non-robust or functional sensitizable off-input. The goal is to minimize the number of FS off-inputs. Thus, an attempt is made to generate a test under which the largest number of FS candidate off-inputs is either robust or non-robust. Because the transition on a non-robust off-input can mask the transition on the corresponding on-input while this can never happen in the case of a robust off-input, it is preferred to have as large a number of robust off-inputs as possible. The FS off-input candidates are processed one at a time. For each candidate, the first attempt is to convert it into a robust off-input by assigning a stable non-controlling value to it. Similar to the algorithm in Section 7.2, the forward implication and the backward justification processes are used to check if the desired transition can be assigned. If the attempt to convert an FS candidate off-input into a robust off-input fails, its value is left unspecified and the algorithm proceeds with the next candidate in the list.

STEP 3: *Try converting the FS off-inputs that have their values still unspecified into non-robust off-inputs one at a time.*

After all FS off-input candidates have been processed, the algorithm continues with the ones that have their values still unspecified and tries to convert them into non-robust off-inputs by assigning a cv→ncv transition to them.

STEP 4: *Assign* ncv→cv *transition to FS off-inputs that have their values still unspecified.*

If the attempt to convert some FS candidate off-input into non-robust off-input fails, a ncv→cv transition is assigned to it, i.e., it is converted into an FS off-input.

The order of processing the FS candidate off-inputs is very important for generating good quality FS tests. Therefore, the algorithm uses the partial timing information available under the current set of mandatory assignments and their implications to calculate the earliest arrival time of the transition on each FS candidate off-input. This information is then used for calculating the slack of each candidate. The candidates with the larger slack are processed first since they are less likely to allow the propagation of the defect to the primary output (see example in Figure 7.9). Therefore, it is desired that they be converted into robust or non-robust off-inputs. The FS candidate off-inputs with a smaller slack have a higher chance of sensitizing the given FS path under faulty conditions and they can be converted into FS off-inputs. Each time after a FS candidate off-input has successfully been assigned a transition the set of mandatory assignments and their implications are updated and the slacks of the FS off-input candidates are recomputed.

STEP 5: *Justification.*

After the appropriate transitions have been assigned to all FS candidate off-inputs, the final justification process is done in a way that maximizes the chance that the non-robust off-inputs arrive as early as possible while the FS off-inputs arrive as late as possible. If the target path has both, non-robust and FS off-inputs, and if the fanin cones of some non-robust and FS off-inputs intersect, the justification process is done so that the non-robust off-input arrives as early as possible.

To generate tests under which the FS off-inputs arrive as late as possible, the fanins to a gate whose output needs to be justified are sorted such that the latest arriving signal is the first in the list. If the transition that needs to be justified is cv→ncv, an attempt is made to assign this transition to all fanins that could have a transition under the current, partially assigned vector pair, processing them as they appear in the fanin list. If the transition that needs to be justified is ncv→cv, the algorithm tries to assign this transition to the fanin with the latest arrival time among all the fanins that could have a transition under the partially assigned test and a stable non-controlling value to the rest of the inputs. If there is a conflict in this justification process,

118   CHAPTER 7

the algorithm backtracks to the last decision point and attempts the justification with the next fanin in the list. In order to achieve that the non-robust off-input arrives as early as possible a similar strategy is used but the fanins are ordered such that the earliest arriving signal is first in the list.

STEP 6: *Specify the values of primary inputs that are still unspecified such that the number of transitions at primary inputs is minimized.*

A summary of the algorithm is given in Figure 7.10. This algorithm cannot guarantee that a fault on an FS path will be detected. Detection can be guaranteed only if all primitive faults have been tested.

For each functional sensitizable path repeat:

1. Assign desired transitions at on-inputs.

2. Convert FS candidate off-inputs into robust off-inputs (if possible), one at a time. Leave the candidate off-inputs, which could not be converted into robust off-inputs, unassigned. Timing information is used to determine the order of processing of the candidates.

3. Convert unassigned off-inputs into non-robust off-inputs (if possible), one at a time. Timing information is used to determine the order of processing of the candidates.

4. Convert the rest of unassigned off-inputs to FS off-inputs.

5. Do backward justification. Timing information is used to make decisions during the justification process.

6. Assign values to the remaining primary inputs that are still unassigned.

Figure 7.10.   Summary of the algorithm for generating FS tests.

7.5   TESTS FOR PRIMITIVE FAULTS

The primitive faults have been defined in Section 5.3. Since testing of the primitive faults of cardinality 1 has been addressed in Sections 7.1, 7.2

and 7.3, this section focuses on testing primitive faults with two or more paths. Therefore, in the sequel, the term "primitive fault" will be used to denote the primitive faults of cardinality higher than 1.

Testing primitive faults was first addressed by Ke and Menon [78]. They propose a technique for synthesizing a two-level circuit such that all path delay faults in the circuit are primitive. Since representing a circuit in a two-level form may not be practical, algebraic transformations are used to convert a two-level circuit into a multi-level circuit. The test set generated for the two-level circuit can also detect all primitive faults in the multi-level circuit. However, multi-level circuits obtained using only algebraic transformations are not area-efficient and the synthesis tools usually use various Boolean transformations for area, performance and power optimizations [9].

A method for identifying and testing primitive faults in multi-level circuits ha been reported by Krstić et al. [87]. It is reviewed in this section. Recently, two other methods for testing primitive faults in multi-level circuits were presented by Sivaraman and Strojwas [137, 138] and Tekumalla and Menon [143, 144]. Sivaraman and Strojwas [137, 138] identify primitive faults by identifying sets of paths that determine the signal stabilization time at the circuit outputs. To be able to handle medium sized benchmark circuits, an iterative approach is proposed such that finding primitive faults of cardinality n requires previous identification of all primitive faults of cardinality $(n-1)$. For larger circuits, due to the memory constraints only low cardinality (up to 3) primitive faults are possible to be identified. Tekumalla and Menon propose techniques for identifying primitive faults in combinational [143] and sequential circuits [144]. The concept of sensitizing cubes is used to reduce the search space for primitive faults. The procedure can also identify faults that cannot be part of any primitive fault and do not need to be tested.

All individual paths in primitive faults of cardinality higher than 1 are functional sensitizable paths. The methodology proposed by Krstić et al. [87] identifies primitive faults by considering individual FS paths and by identifying all primitive faults that the given FS path is involved in. The algorithm is based on the observation that single path delay faults contained in a primitive fault have to merge at one or more gates to form a multiple path delay fault and such gates can be quickly identified. The technique can be combined with timing analysis or any other method for selecting functional sensitizable paths that need to be tested (see Section 5.2).

The following notation and definitions will be used in the description of the algorithm for identifying and testing primitive faults. For a given signal s and a vector v_2, two kinds of FS paths containing signal s and terminating at the same primary output can be defined: (1) FS paths for which the signal s assumes a controlling value under vector v_2 (**cv-FS paths through s**) and (2) FS paths for which signal s assumes a non-controlling value under vector v_2 (**ncv-FS paths through s**). For simplicity, the main ideas of the proposed methodology are explained assuming that the circuit contains only 2-input gates. Extension to circuits containing gates with an arbitrary number of inputs is straightforward. Multiple output circuits are handled by separately processing each primary output and its fanin cone.

7.5.1 Co-sensitizing gates

A primitive fault must have at least one merging gate. If one identifies gates that can never be the merging gates for any primitive fault, then one can significantly reduce the effort required to identify and test primitive faults. This is especially true if the number of non-merging gates represents a significant fraction of the total number of gates in the circuit. Gates that could *possibly* be the merging gates in some primitive fault are called **co-sensitizing gates**. Gates on which no merging of FS paths to form a primitive fault is possible are called **non-co-sensitizing gates**. As will be shown later, identification of co-sensitizing gates is easier than the identification of merging gates.

Co-sensitizing gates with respect to a primary output. Since single paths in a primitive fault end at the same primary output, the identification of primitive faults can be done by separately considering each primary output and its fanin cone. Therefore, co-sensitizing gates can be defined with respect to a primary output.

DEFINITION 7.7 *A 2-input gate g is a **co-sensitizing gate with respect to a primary output** if every input to the gate has at least one cv-FS path passing through it.*

A gate with more than two inputs is a co-sensitizing gate if it has at least two inputs with non-zero number of cv-FS paths passing through them. If an input to some gate has no cv-FS paths passing through it, then all paths passing through that input with a controlling value under vector

v_2 are either robust, non-robust or functional redundant. Therefore, the input cannot be a part of any path in a primitive fault.

EXAMPLE 7.9 Consider the circuit in Figure 7.11. There are six functional sensitizable paths: $P_1 = \{\downarrow, aceghi\}$, $P_2 = \{\downarrow, dghi\}$, $P_3 = \{\downarrow, acfhi\}$, $P_4 = \{\uparrow, acfhi\}$, $P_5 = \{\uparrow, bcfhi\}$ and $P_6 = \{\downarrow, bceghi\}$. The circled number adjacent to a signal shows the number of cv-FS paths through that signal. Gates f, h and i all have one

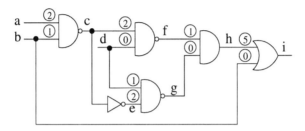

Figure 7.11. Co-sensitizing gates for a given PO.

input with no cv-FS paths. Therefore, they cannot be co-sensitizing gates with respect to the primary output. The only co-sensitizing gates with respect to the primary output are c and g.

If a circuit has multiple outputs, then a gate can be a co-sensitizing gate with respect to one PO and a non-co-sensitizing gate with respect to another PO.

Co-sensitizing gates with respect to an FS path. Condition given by Definition 7.7 identifies the co-sensitizing gates for any FS path in the fanin cone of a given primary output. To find the primitive faults that a given FS path is involved in, the information about co-sensitizing gates can be further refined. In this refinement process, some of the co-sensitizing gates with respect to a given PO might become non-co-sensitizing with respect to the target path. However, all non-co-sensitizing gates with respect to the given PO will stay non-co-sensitizing with respect to any target FS path ending at that PO. Note that a co-sensitizing gate with respect to a given FS path does not have to belong to the path.

DEFINITION 7.8 A co-sensitizing gate g with respect to a primary output o is said to be a **co-sensitizing gate with respect to an FS path** P ending at o if under the set of mandatory assignments (SMA) for P

the following conditions are satisfied: (1) the on-input and off-input of g can be assigned a ncv→cv transition and (2) if gate g is not on P, then there exists a gate $h \in P$ such that h is a co-sensitizing gate with respect to P and g is in the fanin cone of gate h.

Following examples illustrate the conditions of Definition 7.8.

EXAMPLE 7.10 Consider the circuit in Figure 7.12(a). Gates c and g are co-sensitizing gates with respect to the primary output i. Let the FS path for which we are trying to find the primitive faults be $P_4 = \{\uparrow, acfhi\}$. Under the SMA for P_4 the on-input

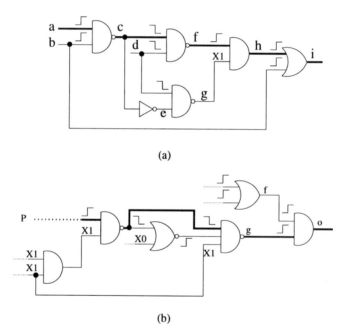

Figure 7.12. Co-sensitizing gates with respect to a given FS path.

a and off-input b to gate c cannot be assigned ncv→cv transition. Therefore, gate c is not a co-sensitizing gate with respect to path P_4. Also, since both inputs to gate g cannot be assigned a ncv→cv transitions under the SMA for P_4, gate g is not a co-sensitizing gate with respect to P_4.

EXAMPLE 7.11 Assume that the logic in Figure 7.12(b) is a part of some larger circuit. Let path P be the target FS path and let gates g and f be the co-sensitizing gates with respect to the primary output o. The SMA for path P is shown in the figure. Under this SMA both inputs to gate f can be assigned ncv→cv transitions. However, gate g is the last co-sensitizing gate in P (closest to the primary output). Therefore, no gate outside the fanin cone of gate g can be a co-sensitizing gate for path P and gate f cannot be a co-sensitizing gate for path P.

7.5.2 Merging gates

There are several reasons why a co-sensitizing gate may not be a merging gate. First, computing the exact number of cv-FS paths through each signal in the circuit can be prohibitively expensive since it requires consideration of two vectors (v_1 and v_2). Approximate values for the number of cv-FS paths through each signal can be efficiently computed by using only the information about the second vector v_2 (see Section 5.2). However, these numbers are pessimistic in the sense that a functional redundant path may be counted as an FS path. On the other hand, FS paths can never be misclassified. Second, even if the cv-FS path count can be computed exactly, the information about the number of cv-FS paths passing through the inputs to a co-sensitizing gate does not take into account any correlations between these paths. Correlations between the FS paths can be twofold: the correlation manifested by a co-sensitizing gate for which the cv-FS paths cannot pass through together under any input vector pair and the correlation manifested by a co-sensitizing gate for which a multiple delay fault involving only FS paths exists, but it is a superset of some primitive fault. One simple way to account for some of the correlations between the paths passing through a given co-sensitizing gate is to assign ncv→cv transition to both of its inputs and find the forward and backward implications of those assignments. If a conflict is detected, then the gate is not a co-sensitizing gate.

7.5.3 Identifying FS paths not involved in any primitive fault

As discussed in Section 5.2, not all single path delay faults need to be selected for testing. This is because not all faults participate in some primitive fault. If all robustly and non-robustly testable path delay faults are selected, then some FS paths do not have to be selected. In addition to the techniques described in Section 5.2 for identifying the set of FS paths that need to be tested, the information about co-sensitizing

124   CHAPTER 7

gates can also be used to perform the selection. If on FS path P there are no co-sensitizing gates with respect to P, then path P does not have to be tested.

EXAMPLE 7.12 Consider again the circuit in Figure 7.12(a). Gates c and g are co-sensitizing gates with respect to the primary output. The SMA and their implications for the functional sensitizable path $P_4 = \{\uparrow, acfhi\}$ are as shown. Note that in the target path there are no co-sensitizing gates with respect to the primary output (see Example 7.10). Therefore, the target path does not have to be tested. The fault on the target path can be observed only if paths $\{\downarrow, dfhi\}$ and $\{\uparrow, bi\}$ are also faulty. However, both of these paths are robustly testable and if any of them is faulty the robust test set will detect the fault. Similarly, it can be determined that path $P_5 = \{\uparrow, bcfhi\}$ does not have co-sensitizing gates and it does not have to be tested.

Information about FS paths that do not require testing can be used to update the number of cv-FS paths through each signal.

EXAMPLE 7.13 Consider the circuit in Figure 7.12(a). Paths P_4 and P_5 pass through signals c and h with controlling values but these paths do not have to be tested. Therefore, the number of cv-FS paths through signals c and h reduces by 2. The updated values for cv-FS paths are shown as circled numbers adjacent to the signals in Figure 7.13.

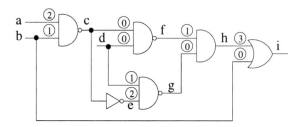

Figure 7.13.   Updating the number of cv-FS paths.

The process of eliminating FS paths that do not have co-sensitizing gates and updating the number of cv-FS paths through each signal can be repeated through several iterations. Updating the values for cv-FS paths results in a smaller number of co-sensitizing gates and a smaller number of FS paths that have to be considered for test generation for primitive faults. However, there exist FS paths that do not have to be tested but cannot be identified by examining only the number of co-sensitizing gates. This is because not every co-sensitizing gate is a merging gate.

EXAMPLE 7.14 Consider the circuit in Figure 7.13. Gate c is a co-sensitizing gate with respect to the FS path $P_3 = \{\downarrow, acfhi\}$. Therefore, this path cannot be eliminated from testing based on the number of co-sensitizing gates. However, it can be shown that path P_3 can be co-sensitized with a robust path $\{\downarrow, bcfhi\}$ and it need not be tested.

7.5.4 Algorithm for identifying and testing primitive faults of cardinality 2

A given FS path can participate in many primitive faults. The cardinality k of these primitive faults can be anywhere from 2 up to the total number of FS paths terminating at the given PO. To guarantee that a fault on an FS path will not affect the performance of the circuit, all primitive faults that involve the FS path have to be identified and tested. The probability that k long functional sensitizable paths will simultaneously be affected by defects significantly reduces as k increases. Therefore, identification and testing of primitive faults of low cardinality is of the highest significance. The following algorithm focuses on primitive faults of cardinality 2.

The algorithm for identifying and testing primitive faults of cardinality 2 consists of two parts [87]. First, the co-sensitizing gates with respect to the given PO are identified. Second, the FS paths that participate in primitive faults of cardinality 2 and all primitive faults of cardinality 2 that include the given FS path are identified. The flowchart of the algorithm is given in Figure 7.14.

Part I: Identifying co-sensitizing gates. This part of the algorithm involves three steps.

STEP 1: *Identify co-sensitizing gates with respect to the given PO.*

This requires knowledge about the number of FS paths passing through each signal. One way to find this number is by using the technique described in Section 5.2.1. That technique gives the upper bound on

126   CHAPTER 7

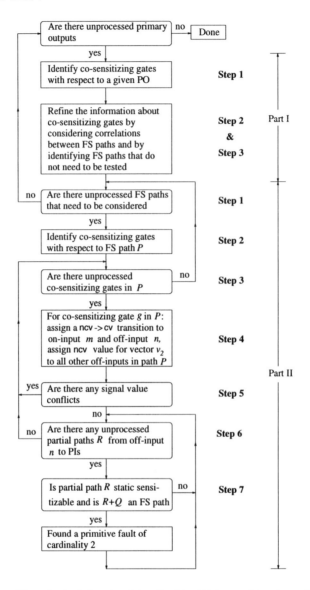

Figure 7.14. Flowchart of the algorithm for identifying and testing primitive faults of cardinality 2.

the number of cv-FS and ncv-FS paths through each signal. Therefore, the number of co-sensitizing gates we identify is also an upper bound.

STEP 2: *Account for some of the correlations between the cv-FS paths that pass through a given co-sensitizing gate.*

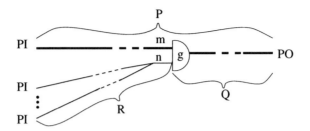

Figure 7.15. Testing path P.

This is done by assigning a ncv→cv transition to both inputs of the co-sensitizing gate and by checking for conflicts during the implication process (see Section 7.5.3).

STEP 3: *Update the number of cv-FS and ncv-FS paths through each signal by identifying FS paths that do not have any co-sensitizing gates under the set of mandatory assignments.*

This step further refines the set of co-sensitizing gates with respect to the primary output (see Section 7.5.3).

Part II: Identifying primitive faults of cardinality 2 that include given FS path. Circuit in Figure 7.15 will be used to explain the second part of the algorithm. Gate g is a co-sensitizing gate for the FS path P. The off-input n of g is an FS off-input. Partial paths R represent possible paths associated with the FS off-input n (paths that can propagate a transition from PI to gate g). Also, Q denotes the partial path of P from gate g to the primary output. The algorithm for finding the primitive faults of cardinality 2 for FS path P involves several steps. Note that whenever the algorithm needs to return to some previous step, all signal values that have originally existed at that step have to be restored.

STEP 1: *Check if there are any FS paths that need to be considered for primitive fault identification. If yes, go to Step 2. If not, continue by processing the next primary output and its fanin cone.*

Functional sensitizable paths that need to be considered are the ones that have at least one co-sensitizing gate with respect to the primary output.

128   CHAPTER 7

STEP 2: *Identify co-sensitizing gates with respect to target FS path P, i.e., the co-sensitizing gates under the SMA for P.*

STEP 3: *Check if there are any unprocessed co-sensitizing gates with respect to P on path P. If yes, go to step 4. If not, go to Step 1.*

STEP 4: *For co-sensitizing gate $g \in P$ assign:*

(i) *a ncv→cv transition to the on-input m and to the off-input n of the co-sensitizing gate g, and*

(ii) *a non-controlling value for the second vector v_2 to off-inputs of all other co-sensitizing gates in P.*

This limits the cardinality of primitive faults to 2 and results in a new set of mandatory assignments (SMA') for path P.

STEP 5: *Check if SMA' is consistent. If yes, go to Step 6. If not, go to Step 3.*

STEP 6: *Check if there is any unprocessed partial path R from the off-input n to any PI. If yes, go to Step 7. If not, go to Step 3.*

To find a co-sensitizing path that together with P forms a primitive fault of cardinality 2, the partial path R must be statically sensitizable under SMA' and it also must be contained in the FS path obtained by concatenation of partial paths R and Q (denoted $R+Q$). The strategy presented in [32] is used to quickly identify signals in the input cone of signal n for which, under the current SMA, no static sensitizable path can pass through. The information about the number of cv-FS and ncv-FS paths passing through each signal is used to prune the search for the partial path (this will be illustrated in Example 7.15).

STEP 7: *Check if a statically sensitizable partial path R to PI is found and path $R + Q$ is functional sensitizable. If yes, a primitive fault of cardinality 2 is found. Go to Step 6.*

The following example illustrates the second part of the algorithm.

EXAMPLE 7.15 Consider the circuit in Figure 7.16(a). The circled numbers show the number of cv-FS paths passing through signals while the numbers inside boxes show the number of ncv-FS paths. The values are shown only for those signals for which the number of cv- or ncv-FS paths is non-zero. Gates i, o and v are the

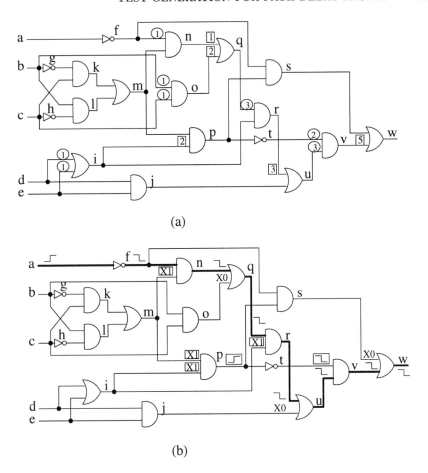

Figure 7.16. Testing primitive faults of cardinality 2.

co-sensitizing gates with respect to the primary output w. Let the target path P for which all primitive faults of cardinality 2 are needed be $\{\uparrow, afnqruvw\}$. This path is shown in bold in Figure 7.16(b). The SMA for P is also shown in the figure (ignore for now the values shown inside the boxes since they are not part of the SMA for P). The only co-sensitizing gate on P is gate v. Therefore, to find a co-sensitized path which together with path P forms a primitive fault of cardinality 2, only partial paths associated with off-input t have to be explored. Note that, if the information about co-sensitizing gates was not used, all partial paths associated with off-inputs i, m and t (potential FS off-inputs in the target path) would have to be explored. Next, a non-controlling value is assigned for the second vector v_2 to all off-inputs other than off-input t and a ncv→cv transition to off-input t. These values are shown within small boxes in Figure 7.16(b). Then, the partial paths from signal t to any primary input are examined. The partial path associated with off-input t must be static sensitizable under the current SMA. Since the path co-sensitized with P has to be functional sensitizable, the search effort can be reduced by using the information about the number of cv-FS and ncv-FS paths passing through each signal. For example, there are no ncv-FS paths through input m of gate p (Figure 7.16(a)) and the gate p in Figure 7.16(b) assumes value 1 for vector v_2. Therefore, all partial paths passing through signal m can be eliminated from further consideration. On the other hand, input i to gate p has 2 ncv-FS paths passing through it and the algorithm has to continue checking all paths passing through signal i. Once a partial path is identified, it is necessary to check if the path co-sensitized with P is functional sensitizable. In this example there are two primitive faults of cardinality 2 containing path P. The first primitive fault includes path P and $\{\uparrow, diptvw\}$. The second primitive fault includes paths P and $\{\uparrow, eiptvw\}$.

7.6 SUMMARY

Experimental results have shown that test sets generated under robust testability criteria cannot detect all timing defects. A defect on a non-robust target path will not be detected if the transitions on certain signals outside the target path arrive late. Among all possible NR tests for a non-robustly testable path some NR tests are better than others in detecting delay defects. A good NR test can tolerate larger timing variations and its probability of being invalidated is smaller. Generating such tests requires the use of circuit timing information.

Generating tests for robust and non-robust path delay faults is not sufficient to cover all possible delay defects. Functional sensitizable paths can, under certain conditions, be responsible for the circuit's performance degradation. Defects on these paths can be detected only if multiple delay faults exist.

Ignoring multiple path delay faults, called primitive faults, in the process of test generation can lead to a poor delay test quality. It is very difficult to identify and test all primitive faults in multi-level circuits. The existing techniques can efficiently deal only with primitive faults that consist of a small number of single paths.

8 DESIGN FOR DELAY FAULT TESTABILITY

This chapter presents design techniques to improve the delay fault testability. Since in many designs, the coverage for path delay faults is unacceptably low, most of the research in this area has concentrated on improving the path delay fault testability. Path delay fault testability can be defined with respect to several factors: the number of faults to be tested, the number of tests that need to be applied to test all path delay faults, the number of faults that can be guaranteed to be detected independent of delays outside the target path, etc. This chapter describes design for testability techniques such as test point insertion and use of partial scan as well as techniques for resynthesizing the circuit such that its path delay fault testability is improved.

8.1 IMPROVING THE PATH DELAY FAULT TESTABILITY BY REDUCING THE NUMBER OF FAULTS

Many circuits have a very large number of paths and the inability to efficiently test all of them is a serious limitation of the path delay fault model. As suggested in Chapter 3, testing can target only the longest path delay faults. This strategy, however, would not help the testing

132   CHAPTER 8

of performance optimized designs in which there could be a very large number of long paths [113]. The problem of having to test a large number of faults can be alleviated by reducing the total number of paths in the circuit. Reduction of the total number of paths can also help in decreasing the time required by the timing analysis tools since that time is, in general, proportional to the number of paths in the circuit. The following example illustrates how path count reduction can increase path delay fault testability when test generation is affordable for a subset of path delay faults.

EXAMPLE 8.1 Let the path distribution for a design be as shown in Figure 8.1 [85]. Horizontal axis represents the path length relative to the longest sensitizable path delay in the circuit. Vertical axis shows the number of sensitizable paths whose delay is *longer* than the delay shown on the horizontal axis. Let curve (a) represent the path distribution in the original circuit and curve (b) represent the path distribution in the circuit after application of a path reduction technique. Assume that it is possible to test 5,000 paths for delay defects. Selecting 5,000 paths in the original circuit means that all paths longer than 75% of longest sensitizable path delay can be checked for delay faults. On the other hand, selecting 5,000 paths in the circuit with a reduced number of paths means that all paths longer than 59% of the critical path delay can be checked. Therefore, the tests derived for the circuit with reduced path count can offer a better guarantee of the product performance.

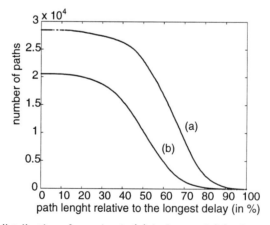

Figure 8.1.   Path distributions for a circuit (a) before and (b) after application of path count reduction technique.

To reduce the path count, it is useful to know the number of paths passing through each signal in the design. The number of paths passing through a signal s can be computed as $NPI(s) \times NPO(s)$, where $NPI(s)$ is the number of partial paths from any primary input to signal s and $NPO(s)$ is the number of partial paths from signal s to any primary output. Values $NPI(s)$ and $NPO(s)$ can be computed using the procedure for computing the number of paths in the circuit (see Section 6.3.2). For computing the NPI labels the circuit is traversed in a topological order from PIs towards POs. For computing the NPO labels the circuit is processed in a topological order from POs towards PIs.

EXAMPLE 8.2 Consider the circuit in Figure 8.2. Each signal s is assigned two labels, $NPI(s)$ and $NPO(s)$. For example, there are 6 partial paths that start at primary inputs and end at the output of gate g_7, and there are 2 partial paths that start at the output of gate g_7 and end at the primary outputs. Therefore, there is a total of $6 \times 2 = 12$ paths passing through the output of gate g_7.

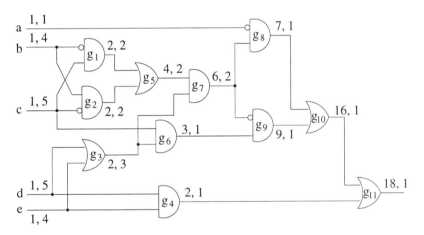

Figure 8.2. Computing the number of paths through each signal.

Several techniques have been proposed for increasing the circuit path delay fault testability by reducing the path count [86, 122, 125]. Pomeranz and Reddy [122] propose a design for testability technique based on test points. The test points are assumed to provide both controllability and observability [1]. A test point at signal s will result in signal s becoming a primary output for the logic in the fanin cone of signal s in the original circuit and a primary input for the logic in the fanout cone of signal s

in the original circuit. The main idea of test point insertion for reducing the path count can be illustrated by the following example.

EXAMPLE 8.3 Consider the circuit in Figure 8.2. Before inserting

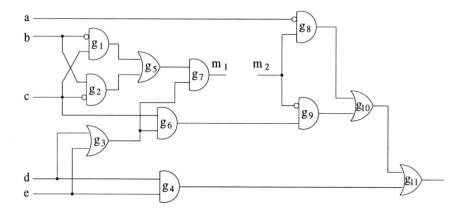

Figure 8.3. Inserting test points to reduce the number of path delay faults.

a test point at the output of gate g_7, the total number of paths passing through this signal was $6 \times 2 = 12$. After placing a test point at the output of gate g_7, the circuit in the test mode is shown in Figure 8.3. The new primary output is denoted m_1 and the new primary input is denoted m_2. In this circuit there are 6 paths that end at signal m_1 and 2 paths that start at signal m_2. Therefore, the number of paths through signals m_1 and m_2 is $6 + 2 = 8$. The total number of paths in the circuit has been reduced by $12 - 8 = 4$ paths.

Addition of a test point divides the paths in the original circuit into subpaths that can be tested through the test points. However, the delays of these subpaths are shorter than the delays of the paths in the original circuit. This means that in addition to the area and performance overhead from the added test points, this technique also requires multiple clock periods for testing the circuit. Use of multiple observation times for testing gate delay faults has been proposed in [72, 106]. Addition of test points can also help in increasing the robust testability of the circuit. This is because a test point can divide a robustly untestable path in the original design into subpaths each of which can be robustly tested. Reducing the number of paths might also reduce the number of tests required to detect all path delay faults [122].

An alternative technique for reducing the path count has been proposed [125]. This resynthesis technique uses the concept of comparison blocks D_i to replace subcircuits C_i of the original design whenever C_i and D_i have the same functionality and the number of paths and/or gates in D_i is smaller than in C_i. The concept of comparison blocks can be illustrated with the following example [125].

EXAMPLE 8.4 Consider function $f_1(x_1, x_2, x_3)$ that is equal to 1 for $x_1 x_2 x_3 \in \{110, 001, 011\}$ which is in decimal form equivalent to $\{6, 1, 3\}$. If variables x_1, x_2 and x_3 are replaced by variables $y_1 = x_3$, $y_2 = x_1$ and $y_3 = x_2$ the function $f_2(y_1, y_2, y_3)$ is equal to 1 for $y_1 y_2 y_3 \in \{011, 100, 101\}$ which is in decimal form equivalent to $\{3, 4, 5\}$. Function f_2 can be implemented using two comparison blocks: a comparison block of type $\geq L$ (where $L = 3$) and a comparison block of type $\leq U$ (where $U = 5$) as shown in Figure 8.4. An implementation of $\geq L$ comparison block for $L = (l_1 l_2 \ldots l_n)$ is

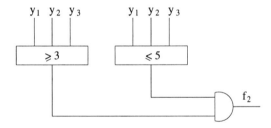

Figure 8.4. Implementation of f_2 using comparison blocks.

given in Figure 8.5(a). Value l_1 denotes the most significant bit of L while l_n denotes the least significant bit of L. The types of gates

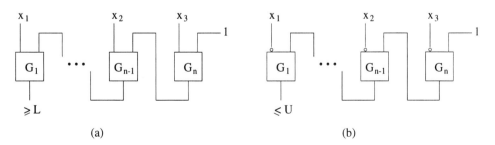

Figure 8.5. Implementation of (a) $\geq L$ and (b) $\leq U$ comparison blocks.

G_i are determined from:

$$G_i = \begin{cases} \text{AND} & \text{if } l_i = 1 \\ \text{OR} & \text{if } l_i = 0 \end{cases}$$

Gate G_n is replaced by a direct connection to x_n when $l_n = 1$ and by a constant 1 when $l_n = 0$. Similarly, an implementation of $\leq U$ comparison block for $U = (u_1 u_2 \ldots u_n)$ is given in Figure 8.5(b). Value u_1 denotes the most significant bit of U while u_n denotes the least significant bit of U. In this case the types of gates G_i are found as:

$$G_i = \begin{cases} \text{AND} & \text{if } u_i = 0 \\ \text{OR} & \text{if } u_i = 1 \end{cases}$$

Gate G_n is replaced by an inverter driven by x_n for $u_n = 0$ and by a constant 1 when $u_n = 1$. Therefore, the implementation of function $f_2(y_1, y_2, y_3)$ from Figure 8.4 using the comparison blocks ≥ 3 and ≤ 5 is given in Figure 8.6.

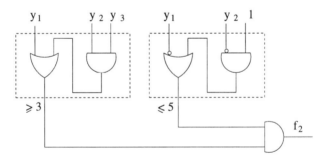

Figure 8.6. Implementation of f_2 using comparison blocks.

From any input variable x_i to the output of a function implemented using comparison blocks there can be at most two paths. Therefore, the comparison blocks can be used to replace logic that has larger number of paths. Finding a representation of a given function that can be implemented using comparison blocks requires considering permutations of the function variables and checking if the minterms (for which the function is equal to 1) belong to a set of consecutive decimal numbers. To limit the increase of the circuit depth and the time for finding comparison blocks, the number of inputs to the comparison blocks has to be kept small.

Krstić et al. [86] propose a path reduction method based on redundancy addition and removal procedure RAMBO [44]. This procedure uses ATPG techniques to decide where to add and where to remove redundancy in order to perturb and simplify a given network during network optimization [44]. The principle can be illustrated with the following example.

EXAMPLE 8.5 The circuit in Figure 8.7(a) is irredundant. In order to simplify the circuit, it can be first perturbed by adding connection $g_5 \rightarrow g_9$. The added signal will not change the circuit func-

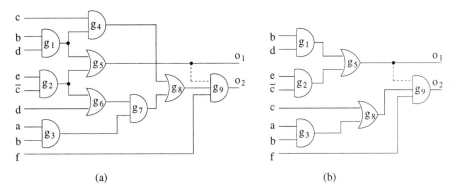

Figure 8.7. Redundancy addition and removal.

tion since it is redundant. However, the addition of connection $g_5 \rightarrow g_9$ will cause two previously irredundant connections, $g_1 \rightarrow g_4$ and $g_6 \rightarrow g_7$, to become redundant and thus removable. The simplified circuit is given in Figure 8.7(b). RAMBO can efficiently identify connections that would create more redundancies for removal.

The above technique can be generalized to allow addition of gates as well as multiple wires [27]. The summary of the path reduction procedure based on RAMBO is given in Figure 8.8. It consists of four steps.

STEP 1: *For each gate g_i calculate: (i) the number of partial paths starting from any primary input and ending at gate g_i, $NPI(g_i)$, and (ii) the number of partial paths starting from gate g_i and ending at any primary output, $NPO(g_i)$.*

These two numbers for all gates can be obtained in linear time as illustrated in Example 8.2.

---

1. For each gate g_i: find $NPI(g_i)$ and $NPO(g_i)$;

2. For each connection $w_k = g_i \rightarrow g_j$: $cost(w_k) = NPI(g_i) * NPO(g_j)$;

3. For each fault $f_k = (w_k, fault_k)$
   if (f_k is redundant) remove f_k;
   else {
       find a set of candidate connections for addition S;
       For each connection $w_a \in S$
           if (w_a is redundant and $cost(w_k) \geq cost(w_a)$)
               add w_a and remove w_k; }

4. For each gate g_i {
   Find a list of faults L that g_i dominates;
   For each fault $f \in L$
       if (f is redundant) remove f;
   Generate the optimization table;
   if (connection w_c corresponding to column c is redundant)
       For each marked row r in column c
           $cost\_to\_be\_removed = cost\_to\_be\_removed + cost(w_r)$;
       if ($cost(w_c) \leq cost\_to\_be\_removed$) add w_c;
       For each marked row r in column c
           if (fault $f_r = (w_r, fault_r)$ in row r is redundant) {
               remove f_r;
               $removed\_cost = removed\_cost + cost(w_r)$; }
       if ($removed\_cost \leq cost(w_a)$)
           backtrack by adding in all the removed connections w_r and remove w_a; }

Figure 8.8. Path reduction algorithm based on redundancy addition and removal.

STEP 2: *Find the number of paths through each connection in the circuit (referred to as the **cost of a connection**). The cost of connection $g_i \rightarrow g_j$ can be found as $NPI(g_i) \times NPO(g_j)$.*

STEP 3: *Try reducing the path count by removing connections and adding alternative connections. Process connections in the decreasing order of their cost.*

For each connection to be removed (referred to as a *target connection*), a set of candidate connections for addition is identified. This set can be identified by finding the set of mandatory assignments for the target connection [44]. Candidate connections for addition, for a given target connection, are those signals which, if added to the circuit, would cause the set of mandatory assignments for the target connection to become inconsistent. However, in order not to change the function of the circuit, the candidate connection for addition must be redundant. Moreover, to reduce the total number of paths in the circuit, it is required that the number of paths created by adding a given redundant connection is at most equal to the cost of the target connection. The following example illustrates the procedure.

EXAMPLE 8.6 The circuit in Figure 8.9 is irredundant. It has 18 paths and 11 gates. When we try to remove connection $g_7 \to g_9$ we find that, for testing the stuck-at fault ($g_7 \to g_9$, s-a-0), the set of mandatory assignments is: $\{a = 1, b = 0, c = 1, g_1 = 1, g_2 = 0, g_3 = 1, g_4 = 0, g_6 = 1, g_7 = 1, g_8 = 0\}$. This set is shown in Figure 8.9. If connection $b \to g_9$ is added, then

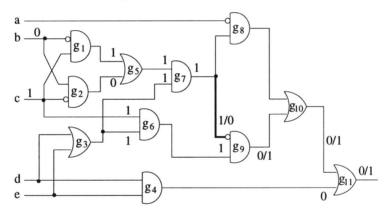

Figure 8.9. Reducing the number of paths in the circuit.

to test the target fault ($g_7 \to g_9$, s-a-0) signal b should be assigned a value 1. However, that is incompatible with value $b=0$ from the above set of mandatory assignments. Therefore, addition of connection $b \to g_9$ would cause the target connection $g_7 \to g_9$ to become redundant. In order not to change the functional behavior of the circuit, before adding the candidate connection $b \to g_9$, it is necessary to check if connection $b \to g_9$ is redundant. Computing the set of mandatory assignments for connection $b \to g_9$ shows that it is redundant. As a last step before performing the redundancy addition and removal, it is necessary to check whether or not the number of paths in the new circuit would be at most equal to the number in the original circuit. Addition of connection $b \to g_9$ would create $NPI(b) \times NPO(g_9) = 1 \times 1 = 1$ path while removal of wire $g_7 \to g_9$ would cause $NPI(g_7) \times NPO(g_9) = 6 \times 1 = 6$ paths to disappear. Therefore, by performing the redundancy addition and removal the total number of paths in this circuit can be reduced from 18 to 13. The resulting circuit is shown in Figure 8.10(a). After the network is perturbed, some con-

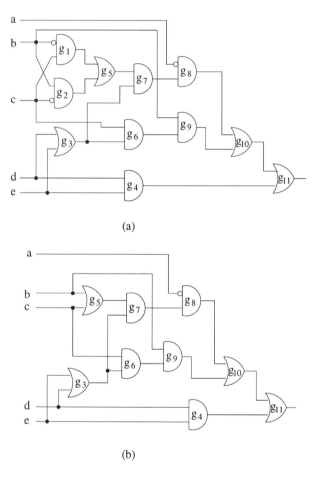

Figure 8.10. Circuit with reduced number of paths.

nections other than the target connection might also become redundant. Therefore, for each target connection before attempting to find candidate connections for addition a check is performed to decide if it is already redundant. In the above circuit, this process identifies connections $b \to g_1$ and $c \to g_2$ as redundant and they can be removed. The new circuit has 11 paths and 9 gates (Figure 8.10(b)).

STEP 4: *Try reducing the path count by adding connections and removing their alternative connections.*

This step is similar to Step 3. The difference is that Step 3 starts with a target fault which should be removed and searches for candidate connections for addition while Step 4 starts from a candidate connection for addition and tries to remove as many connections as possible.

DEFINITION 8.1 A gate g_i is said to be a **dominator** of gate g_j with respect to a primary output o iff all paths from g_j to o pass through gate g_i. If a gate g_i is a dominator of gate g_j with respect to all primary outputs, gate g_i is called an **absolute dominator**.

Dominator and absolute dominator gates [44] are used to guide the search for connections to be added and removed from the circuit. The gates are ordered according to the number of paths that pass through their outputs and the one with the highest number of paths is processed first. For each gate, a list of the faults that this gate dominates is found. Then, all faults in the list are tested and the redundant faults are removed. Next, an optimization table [44] is formed. The columns in this table correspond to the candidate connections for additions. They connect certain nodes (source nodes) identified during redundancy checking with the given dominator gate. All nodes that are not in the fanin or fanout cone of the dominator node are considered as source nodes for candidate connections for addition. The rows in the table correspond to the remaining (irredundant) faults in the fault list. An entry in row r and column c is marked in the table if a given candidate connection for addition corresponding to column c causes the fault in row r to become redundant. Each column in this table will have at least one entry and some columns might have more entries.

EXAMPLE 8.7 Consider the circuit in Figure 8.7(a). The fault list for the absolute dominator gate g_9 is: $(g_8 \rightarrow g_9$, s-a-1), $(f \rightarrow g_9$, s-a-1), $(g_4 \rightarrow g_8$, s-a-0), $(g_7 \rightarrow g_8$, s-a-0), $(c \rightarrow c_4$, s-a-1), $(g_1 \rightarrow g_4$, s-a-1), $(g_6 \rightarrow g_7$, s-a-1), $(g_3 \rightarrow g_7$, s-a-1), $(g_2 \rightarrow g_6$, s-a-0), $(d \rightarrow g_6$, s-a-0), $(a \rightarrow g_3$, s-a-1) and $(b \rightarrow g_3$, s-a-1). The connections in this list belong to the fanin cone of gate g_9 and are not shared with other gates. The only node outside of the input and output cones of node g_9 is g_5. Therefore, the optimization table has two columns, one for the fault ($g_5 \rightarrow g_9$, s-a-0) and another for the fault ($g_5 \rightarrow g_9$, s-a-1). Table 8.1 represents the optimization table for node g_9. Since faults ($g_1 \rightarrow g_4$, s-a-1)

Table 8.1. Optimization table for g_9 in Figure 8.7(a).

	($g_5 \rightarrow g_9$, s-a-0)	($g_5 \rightarrow g_9$, s-a-1)
($g_8 \rightarrow g_9$, s-a-1)		
($f \rightarrow g_9$, s-a-1)		
($g_4 \rightarrow g_8$, s-a-0)		*
($g_7 \rightarrow g_8$, s-a-0)		
($c \rightarrow g_4$, s-a-1)		*
($g_1 \rightarrow g_4$, s-a-1)	*	
($g_6 \rightarrow g_7$, s-a-1)	*	
($g_3 \rightarrow g_7$, s-a-1)		
($g_2 \rightarrow g_6$, s-a-0)		*
($d \rightarrow g_6$, s-a-0)		*
($a \rightarrow g_3$, s-a-1)		
($b \rightarrow g_3$, s-a-1)		

and ($g_6 \rightarrow g_7$, s-a-1) imply a mandatory value 0 at node g_5, they can be made redundant by adding the connection ($g_5 \rightarrow g_9$, s-a-0). Similarly, faults ($g_4 \rightarrow g_8$, s-a-0), ($c \rightarrow g_4$, s-a-1), ($g_2 \rightarrow g_6$, s-a-0) and ($d \rightarrow g_6$, s-a-0) imply a mandatory value 1 at node g_5 and they can be made redundant by adding the connection ($g_5 \rightarrow g_9$, s-a-1). This is the meaning of the marked elements in the optimization table.

For each column in the optimization table, the path reduction that would be obtained if all faults corresponding to the marked rows turn out to be redundant is computed. The candidate connections that correspond to the column with the highest reduction are processed first. After removal of the first fault, the rest of the faults marked for the given column might not be redundant any more. If this happens and the number of paths added is larger than the number of paths removed then the algorithm has to backtrack, i.e., the removed connection is added back to the circuit. The process continues with the next fault in that column. For a given column, the faults that could remove a larger number of paths are processed first.

For some circuits, it might be possible to obtain a larger path count reduction by iterating Steps 3 and 4.

8.2 IMPROVING THE PATH DELAY FAULT TESTABILITY BY INCREASING ROBUST TESTABILITY OF DESIGNS

As discussed in Chapter 5, detection of a delay fault on a robustly testable path is independent of the delays on signals outside the target path. Test generation for robustly testable faults can be efficiently done. Therefore, it is desirable that all or most faults in the circuit can be robustly tested. The reality, however, is that in most designs there is small set of robustly testable path delay faults and a much larger set of non-robust and functional sensitizable faults that can also affect the circuit performance. Detection of all faults on non-robust and functional sensitizable paths requires enormous test generation effort in large circuits and can result in a very large test set (multiple path delay faults need to be tested).

Robust path delay fault testability of a design can be defined as the ratio of the number of robustly testable path delay faults and the total number of faults that need to be tested to guarantee the performance. It is often used as a measure of the path delay testability of a design. Design for testability techniques to improve the robust path delay fault testability of the circuit have been proposed [25, 122, 125, 145].

Chakraborty *et al.* [25] investigate the causes of low robust path delay fault coverage in sequential circuits. The undetectable faults are classified into three categories: (1) faults that cannot be combinationally activated, (2) faults that cannot be sequentially activated and (3) faults that cannot be observed at the primary outputs. Experimental results have shown that most of the undetected faults belong to the second class [25]. Figure 8.11 illustrates the relative sizes of the three groups of faults reported in [25]. Faults in the first category (combinationally non-activated faults) represent faults that cannot be robustly tested when only the combinational portion of the sequential design is considered. This set can be reduced only if the combinational logic is resynthesized or changed. Any fault that is combinationally undetectable is also sequentially undetectable. Faults in the second category (sequentially non-activated faults) represent faults for which it is impossible to find two successive vectors such that the given fault is activated in the sequential design. Faults in the third category (unobservable faults) represent faults that can be activated in the sequential design but the path destination

144 CHAPTER 8

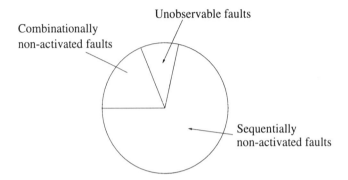

Figure 8.11.   Untestable path delay faults in sequential circuits.

is a flip-flop and the fault effect cannot propagate from this flip-flop to any primary output. The number of faults in the second category can be reduced by using partial scan to improve the controllability of the state lines. Partial scan can also help reduce the number of faults in the third class by improving the observability of the state lines. Chakraborty *et al.* [25] choose the flip-flops to be scanned such that long feedback loops are broken [31]. Also, to facilitate delay testing, the scan flip-flops are assumed to be able to store two bits of state (see Section 2.2.1).

For combinational circuits, robust path delay fault testability can be increased by using the test point insertion technique described in Section 8.1. An improvement of this technique resulting in a smaller number of inserted test points and a smaller number of clock periods to test all the faults that can affect the circuit performance has been published by Uppaluri *et al.* [145]. It is based on using ATPG techniques to decide the locations for test points and on considering only the paths that need to be tested in order to guarantee the circuit performance instead of considering all faults in the design.

8.3 IMPROVING PATH DELAY FAULT TESTABILITY BY INCREASING PRIMITIVE DELAY FAULT TESTABILITY

The number of robustly testable path delay faults is usually very small and design for robust testability techniques can result in a large area/performance overhead. Allowing some of non-robustly testable faults and primitive faults of low cardinality can help reduce the overhead at the expense of test generation for these faults. As discussed in Section 7.5, many circuits have a large number of primitive faults of low cardinality. Tests for these primitive faults can be derived using the techniques described in Section 7.5.4. Krstić et al. [84] propose a DFT technique that inserts a small number of control points such that the circuit, in the test mode, has only primitive faults of cardinality up to 2. After testing these primitive faults, the circuit performance can be guaranteed. The following example illustrates the idea.

EXAMPLE 8.8 Consider the circuit in Figure 8.12(a). The three path delay faults shown in bold, $P_1 = \{\uparrow, cegijl\}$, $P_2 = \{\uparrow, chjl\}$ and $P_3 = \{\downarrow, dfhjl\}$ belong to a primitive fault of cardinality three, $\Pi = \{P_1, P_2, P_3\}$. Gates h and j are the merging gates for Π. It can be shown that Π is the only primitive fault of cardinality higher than 2 in this circuit. To ensure that the circuit is delay fault-free by testing only primitive faults of cardinality up to 2, a control point can be inserted at the fanout of gate i, as shown in Figure 8.12(b). In test mode the test pin m will be assigned a stable

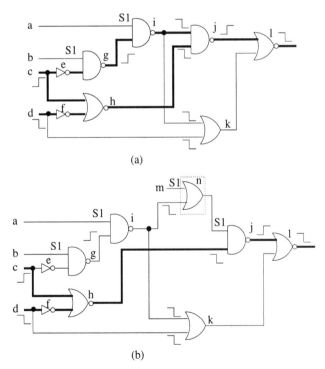

Figure 8.12. Inserting test points to reduce the cardinality of primitive faults.

controlling value S1. In the functional mode the test pin m will be assigned a stable non-controlling value S0. In test mode paths P_2 and P_3 form a primitive fault of cardinality 2, $\Pi_1 = \{P_2, P_3\}$. This fault can be tested using the procedure described in Section 7.5.4. If Π_1 is not faulty, it can be guaranteed that, in the functional mode, the performance of the circuit will not be affected by the primitive fault Π. This is because all three paths P_1, P_2 and P_3 have to be simultaneously faulty for the primitive fault Π to affect the performance of the circuit and it is known that either P_2 or P_3 is fault-free (since Π_1 is fault-free). In this case, addition of a test point reduces the cardinality of the primitive fault from 3 to 2.

8.3.1 Primitive faults of cardinality k > 2

This section considers necessary conditions for existence of a primitive fault of cardinality $k > 2$ for a given FS path P. The following definitions are needed.

DEFINITION 8.2 **Cardinality of a co-sensitizing gate** g, denoted as $C(g)$, is the sum of the number of cv-FS paths passing through the inputs to gate g. The cardinality of a co-sensitizing gate g is $C(g) = 2 + 1 = 3$.

EXAMPLE 8.9 Consider the circuit in Figure 8.13.

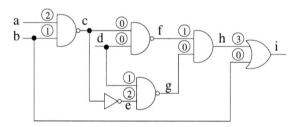

Figure 8.13. Cardinality of a co-sensitizing gate.

Let $cv\_FS(i)$ denote the number of cv-FS paths passing through signal i and let $N_{ub}(g, k)$ denote the **upper bound on the number of primitive faults of cardinality k for which gate g is a merging gate**. Using the information about $cv\_FS(i)$ for each input to a co-sensitizing gate g, $N_{ub}(g, k)$ can be computed by counting the number of different ways for choosing k paths such that not all of them pass through the same input of g.

EXAMPLE 8.10 Let gate g have two inputs, a and b, and let $cv\_FS(a) = 3$ and $cv\_FS(b) = 2$. Then, $N_{ub}(g, 2) = \binom{3}{1} \times \binom{2}{1} = 6$, $N_{ub}(g, 3)$

DESIGN FOR DELAY FAULT TESTABILITY   147

$= \binom{3}{2} \times \binom{2}{1} + \binom{3}{1} \times \binom{2}{2} = 9$, $N_{ub}(g,4) = \binom{3}{3} \times \binom{2}{1} + \binom{3}{2} \times \binom{2}{2} = 5$, $N_{ub}(g,5) = \binom{3}{3} \times \binom{2}{2} = 1$, while the number of possible primitive faults of cardinality 6 or higher is 0.

Number $N_{ub}(g,k)$ represents an upper bound which might not be tight. The reasons are the inability to precisely compute the number of cv-FS paths passing through each signal, the inability to find all FS paths that do not have to be considered to identify primitive faults and ignoring the path correlations (see Section 7.5.2). Also, gate g might participate as a merging gate in some primitive faults of cardinality higher than k. Let $N_{identified}(g, k-1)$ denote the **number of identified primitive faults of cardinality (k-1) such that gate g is a merging gate**. If for a given co-sensitizing gate g and given cardinality k we have $N_{identified}(g, k-1) = N_{ub}(g, k-1)$, then gate g cannot be a co-sensitizing gate for any primitive fault of cardinality higher than $(k-1)$. Therefore, with respect to primitive faults of cardinality higher than $k-1$, gate g is a non-co-sensitizing gate. On the other hand, if $N_{identified}(g, k-1) < N_{ub}(g, k-1)$, then gate g should be considered as a co-sensitizing gate with respect to primitive faults of cardinality k.

DEFINITION 8.3 A gate is said to be a **co-sensitizing gate with respect to primitive faults of cardinality k** if $N_{identified}(g, k-1) < N_{ub}(g, k-1)$.

LEMMA 8.1 A necessary condition for the existence of a primitive fault of cardinality $k > 2$ that includes FS path P is that there exists at least one co-sensitizing gate g in P such that (1) the cardinality of g is higher than $(k-1)$ and (2) gate g is a co-sensitizing gate with respect to primitive faults of cardinality k.

The following example illustrates the conditions from Lemma 8.1 for $k = 3$.

EXAMPLE 8.11 Consider again the circuit in Figure 8.13. Let the target path be $P = \{\downarrow, bceghi\}$. Gates c and g are the co-sensitizing gates with respect to the target path. To check if P might be involved in some primitive fault of cardinality 3, first all primitive faults of cardinality 2 have to be found. For gates c and g we have $N_{ub}(c, 2) = 2$, $N_{ub}(g, 2) = 2$, $C(c) = 3$ and $C(g) = 3$. It can be shown that there is only one primitive fault of cardinality 2 in this circuit and it contains paths {falling, bceghi} and {falling, dghi}.

The only merging gate is g. Therefore, $N_{identified}(c, 2) = 0$ and $N_{identified}(g, 2) = 1$. This means that both gates will be classified as co-sensitizing gates with respect to primitive faults of cardinality 3. Therefore, the condition from Lemma 8.1 is satisfied for path P.

8.3.2 Design for primitive delay fault testability

The algorithm for identifying good locations for inserting control points for primitive delay fault testability [84] is based on the knowledge about co-sensitizing gates (see Section 7.5.1). If every FS path in the circuit has at most one co-sensitizing gate, the highest cardinality of a primitive fault in that circuit would be 2. Adding test points such that this condition is satisfied would be a simple solution to achieve a design in which the highest cardinality of a primitive fault is 2. However, this solution can be sub-optimal. This is because some co-sensitizing gates might not be merging gates for any primitive fault (see Section 7.5.2). Also, some co-sensitizing gates might be merging gates only for primitive faults of cardinality 2.

The DFT algorithm for primitive faults consists of two parts. The flowchart of the algorithm is shown in Figure 8.14. The first part identifies all primitive faults of cardinality 2 and finds the potential locations for inserting test points. Addition of a test point can change the testability of many primitive faults in the circuit. Therefore, the second part of the algorithm consists of several iterations of test point insertion and re-evaluation of the circuit for its primitive delay fault testability.

Part I: Identifying potential locations for test points. This part of the algorithm constructs two lists, P_LIST and TP_LIST. The P_LIST contains a list of FS paths that might be involved in some primitive fault of cardinality higher than 2. The TP_LIST contains a list of potential locations for insertion of control points. This phase of the algorithm is identical to the algorithm for identifying primitive faults of cardinality 2 described in Section 7.5.4 with a few additional steps (Steps 4A, 5A, 7A and 7B). The additional steps are shown in bold in the flowchart in Figure 8.14 and are explained below.

STEP 4A: *Check if the necessary conditions for a co-sensitizing gate to participate in a primitive fault of cardinality higher than 2 are satisfied for the co-sensitizing gate $g \in P$. If yes, go to Step 5. If not, go to Step 3.*

DESIGN FOR DELAY FAULT TESTABILITY 149

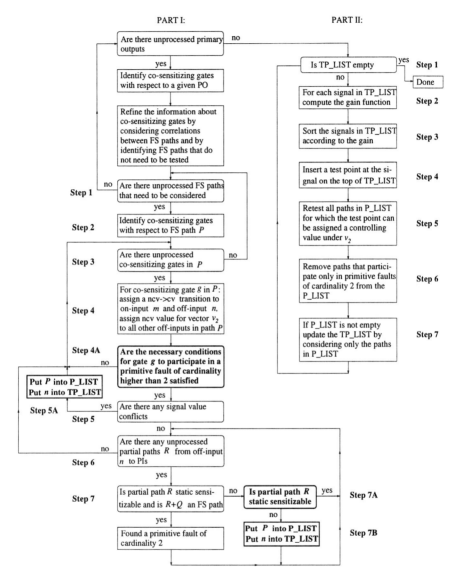

Figure 8.14. Flowchart of the design for primitive delay fault testability algorithm.

Necessary conditions are given by Lemma 8.1. Note that checking whether the conditions of Lemma 8.1 are satisfied requires that all primitive faults of cardinality 2 have been previously identified. This means that the process of identifying primitive faults of cardinality 2 and finding the potential locations for inserting test points cannot be combined into the same pass of the algorithm. The extra pass

150   CHAPTER 8

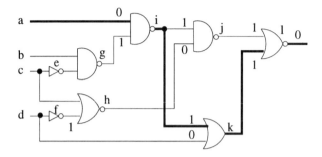

Figure 8.15.   Identifying candidate signals for test point insertion.

can be avoided if the current number of identified primitive faults of cardinality 2 is used instead. However, this might result in a larger number of test points.

STEP 5A: *Insert P into P_LIST and insert signal n into TP_LIST. Store the information that gate g is a co-sensitizing gate for P. Go to Step 3.*

If the set of mandatory assignments after assigning values to the off-inputs of co-sensitizing gates in P (SMA'), as explained in Step 4 of the algorithm in Section 7.5.4, is inconsistent, it means that path P might participate in a primitive fault of cardinality higher than 2.

STEP 7A: *If path $R+Q$ is an FS path and path R is not static sensitizable, go to Step 7B. Else, go to Step 6.*

If the path obtained by concatenation of path R and path Q is an FS path and partial path R is not static sensitizable under SMA', it means that path P might be involved in a primitive fault of cardinality higher than 2.

STEP 7B: *Insert P into P_LIST and insert signal n into TP_LIST. Store the information that gate g is a co-sensitizing gate for P. Go to Step 6.*

In designs with a large number of FS paths, only the longest FS paths can be selected and considered for test point insertion. The following example illustrates .5the first part of the algorithm.

EXAMPLE 8.12 Consider the circuit in Figure 8.15. All 2-input gates in this circuit are co-sensitizing gates with respect to the primary output. The target path is shown in bold and the co-sensitizing gates under the SMA for this path are i, k and l. Figure 8.15 shows the SMA' after assigning a controlling value under vector v_2 to the off-input of gate l and non-controlling values under v_2 to the off-inputs of gates k and i. Since SMA' is consistent we need to explore the partial paths from the off-input of gate l to the primary inputs. The partial paths to be explored are: $R_1 = \{\uparrow, aij\}$, $R_2 = \{\downarrow, bgij\}$, $R_3 = \{\uparrow, cegij\}$, $R_4 = \{\downarrow, dfhj\}$ and $R_5 = \{\uparrow, chj\}$. From the set of mandatory assignments in Figure 8.15 it can be seen that partial paths R_1, R_2 and R_3 cannot be co-sensitized in any way together with the target path P (due to the assignment of a non-controlling value to signal i and a controlling value to signal h under SMA'). Partial path R_4 is static sensitizable under the set of mandatory assignments in Figure 8.15 and it is a part of FS path $P_1 = \{\downarrow, dfhjl\}$. Therefore, fault $\{P, P_1\}$ is a primitive fault of cardinality 2. On the other hand, partial path R_5 is not static sensitizable under the set of mandatory assignments in Figure 8.15. This is due to the fact that both inputs to co-sensitizing gate h have to be assigned a controlling value under SMA'. Therefore, path P might be involved in a primitive fault of cardinality higher than 2 and will be inserted in the P_LIST. Also, the off-input of gate l is a potential location for insertion of a control point. Therefore, the off-input of l is added to the TP_LIST. Next, the same needs to be repeated for the remaining two co-sensitizing gates in path P (gates k and i). After processing gates k and i we find that gate k does not need to be while gate i needs to be inserted into the TP_LIST.

After the first phase of the algorithm the TP_LIST does not include all co-sensitizing gates in the circuit. Only the gates in TP_LIST can be the merging gates for primitive faults of cardinality higher than 2 (these gates are co-sensitizing gates with respect to primitive faults of cardinality higher than 2). In the rest of this section only those co-sensitizing gates included in the TP_LIST will be referred to as "co-sensitizing gates". Each path in P_LIST has at least one co-sensitizing gate.

Part II: Iterative addition of test points and retesting of FS paths in *P_LIST*. Addition of a control point can alter the conditions under which a given FS path can be tested. Therefore, after inserting a control point, the paths in *P_LIST* need to be retested in order to re-evaluate their testability. Hence, the second part of the algorithm consists of insertion of test points and retesting of paths in *P_LIST*. This phase involves several steps.

STEP 1: *Check if TP_LIST is empty. If not, go to Step 2. If yes, the algorithm is done.*

STEP 2: *For each signal $s \in TP\_LIST$ compute the gain reflecting the expected benefit of inserting a test point at signal s.*

For each signal s, two different gain functions are computed: *off_input_gain(s)* and *fanin_gain(s)*.

(i) Function *off_input_gain(s)* reflects the number of paths in *P_LIST* for which signal s is an off-input to some co-sensitizing gate g (these paths form a subset denoted as *P_LIST(s)*). The contribution of each such path to the gain of inserting a test point at signal s depends on the number of co-sensitizing gates in the path. This is because if g is the only co-sensitizing gate for path P, after inserting a control point at s path P can be eliminated from *P_LIST*. On the other hand, if gate g is one of several co-sensitizing gates for path P, addition of a control point at signal s does not guarantee that path P can be removed from *P_LIST*. Let $CG(P)$ denote the set of co-sensitizing gates in P. Then, *off_input_gain(s)* is computed as

$$\mathit{off\_input\_gain}(s) = \sum_{P \in P\_LIST(s)} \frac{1}{|CG(P)|}$$

(ii) Function *fanin_gain(s)* reflects the fact that a control point at signal s might reduce the cardinality of primitive faults for some path $P \in P\_LIST$ even though signal s is not an off-input to some co-sensitizing gate g in P. This can happen if signal s is in the fanin cone of some off-input of a co-sensitizing gate in P. On the other hand, if signal s is in the fanin cone of some off-input in P that must be assigned a non-controlling value under v_2, the control point at signal s cannot be used to reduce the cardinality of primitive faults for path P. Let $CV(s)$ be a set of paths in *P_LIST* for which a test point at signal s can be assigned a controlling value under v_2. Then,

$$\mathit{fanin\_gain}(s) = |CV(s)|$$

STEP 3: *Sort the list of potential locations for test points, TP_LIST.*

The list is sorted in descending order of *off_input_gain*. The value of *fanin_gain* is used to break ties. The remaining ties are broken arbitrarily.

STEP 4: *Insert a control point at the signal on the top of the TP_LIST.*

STEP 5: *Retest paths in P_LIST for which the test pin can be assigned a controlling value under v_2.*

STEP 6: *If a path participates only in primitive faults of cardinality up to 2 remove it from P_LIST. If a path cannot be removed from the list, update the information about co-sensitizing gates for the given path.*

STEP 7: *If P_LIST is not empty, update the TP_LIST considering only paths remaining in P_LIST. Go to Step 1.*

The following example illustrates the gain computation in the second phase of the algorithm.

EXAMPLE 8.13 Consider the circuit in Figure 8.16. After the first phase of the algorithm, let the *P_LIST* contain paths $P_1 = \{\uparrow, cegijl\}$, $P_2 = \{\downarrow, bgijl\}$, $P_3 = \{\downarrow, aikl\}$ and $P_4 = \{\uparrow, dkl\}$. Let the *TP_LIST* contain signals h, e, j and g. Also, let the set of co-sensitizing gates for each of the four paths be $CG(P_1) = \{j\}$, $CG(P_2) = \{g\}$, $CG(P_3) = \{i, l\}$ and $CG(P_4) = \{l\}$.

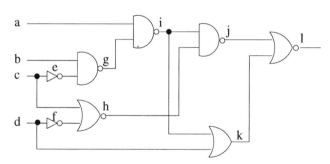

Figure 8.16. Computing the gain.

First, the *off_input_gain* for each signal in *TP_LIST* is computed. Signal h is an off-input to gate j and gate j belongs only to set

$CG(P_1)$. Since $|CG(P_1)| = 1$, we get $\mathit{off\_input\_gain(h)} = 1$. Similarly, we get $\mathit{off\_input\_gain(e)} = 1$. Since signal g is an off-input to gate i and $i \in P_3$, we have $\mathit{off\_input\_gain(g)} = 0.5$. Also, signal j is an off-input to gate l and gate l is present in sets $CG(P_3)$ and $CG(P_4)$. Since $|CG(P_3)| = 2$ and $|CG(P_4)| = 1$, we have $\mathit{off\_input\_gain(j)} = \frac{1}{2} + 1 = 1.5$. Next, the $\mathit{fanin\_gain}$ for each signal in TP_LIST is computed. This gain reflects the number of paths for which the given signal is not an input to a co-sensitizing gate but it is in the fanin cone of some co-sensitizing gate. For example, signal h is in the fanin cone of gate l and gate l is a co-sensitizing gate for 2 paths, i.e., $CV(h) = \{P_3, P_4\}$. Similarly, $CV(e) = \{P_3, P_4\}$, $CV(j) = \{\}$ and $CV(g) = \{P_4\}$. Therefore, we get $\mathit{fanin\_gain(h)} = \mathit{fanin\_gain(e)} = 2$, $\mathit{fanin\_gain(j)} = 0$ and $\mathit{fanin\_gain(g)} = 1$. After computing the gain function, the possible order for signals in TP_LIST is $\{j, h, e, g\}$ or $\{j, e, h, g\}$.

If timing is not considered, addition of test points can adversely affect the performance of the design. To avoid performance degradation, timing driven DFT for primitive delay faults can be considered. A technique similar to that presented by Cheng and Lin [35] can be easily combined with the above approach.

8.4 SUMMARY

Improving the testability of designs rarely comes without additional cost. It usually involves an increase in area, performance degradation, added input/output pins, etc. In delay testing where test generation for guaranteeing the design performance can be computationally very expensive or impossible and where the test set can be extremely large, design for testability techniques offer a viable solution. This is especially true for path delay faults. When it is too expensive to test all faults in the design, design for testability techniques should be combined with the methodologies for selecting the critical path delay faults such that the cost of DFT is reduced by improving only the testability of critical faults while the performance degradation of the design is minimized. Another technique of design for testability involves removal of non-robustly untestable paths. These paths are sometimes referred to as *false paths*. Interestingly, such paths are often associated with redundant stuck-at faults [102], whose identification and removal has been discussed in the literature [36, 73]. However, not every false path has a corresponding redundant stuck-at fault. Even those can be identified by circuit modifications [52]. The advantage of removing false paths is generally a speed up of the circuit, though the modified circuit can be larger.

Recently, there has been interest in the built-in self-test (BIST) design for delay fault testing. BIST for delay faults was initially proposed by Breuer and Nanda [20]. Current directions of work involve generation of patterns, such as single input change patterns or Gray code patterns, that provide high delay fault testability. In addition, circuit modifications like test-point insertion and partial scan have been proposed [110, 115].

9 SYNTHESIS FOR DELAY FAULT TESTABILITY

This chapter focuses on techniques for synthesizing delay fault testable circuits. The presented techniques concentrate on transition or path delay faults. Robust tests for these faults can guarantee their detection independent of signal delays outside the target path. Therefore, it would be ideal if all faults could be tested with robust tests. However, in most designs a significant fraction of delay faults is not robustly testable. Unlike in stuck-at fault testing, redundancy removal techniques cannot be used to eliminate untestable delay faults. There are two reasons for this: (a) An untestable path delay fault is not necessarily redundant and thus it may not be removed. (2) Even if a path is redundant, the signals involved in the path are usually shared by many paths. Removing a path from a circuit would then, require duplication of the logic and thus, result in an area increase. A number of synthesis techniques tuned to obtain circuits with improved robust delay fault testability have been developed [33, 40, 89, 118, 127, 130]. These methods either start from the functional specification of the circuit and synthesize it such that all faults are robustly testable or they identify the robustly untestable faults in the given implementation and modify it to achieve complete robust delay fault testability. Synthesis of robust delay fault testable designs can result in a large area/performance overhead and or additional primary inputs. In case of the path delay faults, more area efficient delay testable designs can be obtained if in addition to robustly testable faults the synthesized designs also allowed to contain validatable non-robustly testable faults as well as primitive faults of cardinality higher than 1. Defects causing all these faults can be detected without regard to circuit delays. However, unless the test set for such a design is obtained during the synthesis process, generating tests for validatable non-robustly testable and primitive faults of cardinality higher than 1 is a complicated task.

Terminology

The following definitions are taken from the literature [40].

A **Boolean function** F of n variables represents a mapping $B^n = \{0,1\}^n \to \{0,1\}$, where B^n is modeled as a binary n-cube. A vertex $v \in B^n$ for which $F(v) = 1$ is a member of the ON-set. Vertex v is also called a **1-vertex**. If $F(v) = 0$, then v is a member of the OFF-set (v is also called **0-vertex**).

A **literal** is a Boolean variable or its complement. A **cube** is a set of literals and in this chapter it will be interpreted as a product of literals. A cube such that every variable of the Boolean function appears in it represents a **minterm**. The minterm can be interpreted as a **vertex** in the n-cube.

A **cover** is a set of cubes and in this chapter it will be interpreted as a sum-of-products. A cube q is said **to cover** the cube r iff q is a subset of r. If a cube q covers only ON-set vertices of a Boolean function F, then q is called an **implicant** of F. A **relatively essential vertex** or an **irredundancy test** of a cube q in a cover C is a vertex that is covered by q and is not contained in any other cube in C.

A **two-level circuit** is a circuit that implements a cover by using an AND gate for each product term and an OR gate for the sum of products, adding inverters on inputs as needed.

Let E be a cover of a two-level circuit C. A set of literals $L = \{l_1, \ldots, l_k\}$ is said to be a **prime literal set** in cube q, $q \in E$, if $L \subseteq q$ and $q - L$ contains a 0-vertex. A set of cubes $Q \subseteq E$ is said to be **redundant cube set** if $E - Q$ contains the same vertices as E, otherwise it is said to be **irredundant**.

A two-level implementation of logic function F is said to be **prime and irredundant** if no literal or implicant of F can be dropped without changing the input/output behavior of the circuit [36].

9.1 SYNTHESIS FOR ROBUST DELAY FAULT TESTABILITY

Delay fault testability is related to the structure rather then the function of the circuit. Most designs optimized for delay and area are not robust delay fault testable. Guaranteeing the temporal correctness of such designs is a difficult problem. A possible solution is synthesis for robust path delay fault testability [33, 40, 89, 118, 127, 130]. Most of the proposed methods for robust delay fault testability are applicable only to combinational/full scan circuits and require extra logic and/or extra inputs. If full scan is assumed, either enhanced or standard scan test application strategy may be used (see Chapter 2).

9.1.1 Combinational and enhanced scan sequential circuits

Synthesis of robust path delay fault (RPDF) testable multi-level functions has been considered by Kundu *et al.* [89]. The proposed synthesis technique is based on decomposing the function according to the Shannon's expansion theorem [82]. Using this theorem, an arbitrary function $F(x_1, x_2, \ldots, x_n)$ can be decomposed with respect to a variable x_i as:

$$F(x_1, x_2, \ldots, x_n) = x_i F_{x_i} + \overline{x}_i F_{\overline{x}_i} \quad (9.1)$$
$$F(x_1, x_2, \ldots, x_n) = (\overline{x}_i + F_{x_i})(x_i + F_{\overline{x}_i}) \quad (9.2)$$

where

$$F_{\overline{x}_i} = F(x_1, x_2, \ldots, x_{i-1}, 0, x_{i+1}, \ldots, x_n)$$
$$F_{x_i} = F(x_1, x_2, \ldots, x_{i-1}, 1, x_{i+1}, \ldots, x_n).$$

Functions $F_{\overline{x}_i}$ and F_{x_i} both depend on at most $(n-1)$ variables. The decomposition given by Equation (9.1) can be implemented using the circuit in Figure 9.1(a) and the decomposition given by Equation (9.2) can be implemented using the circuit in Figure 9.1(b).

If function F is decomposed according to Equation (9.1) or Equation (9.2) and implemented by circuits shown in Figure 9.1(a) or 9.1(b) respectively, and if functions F_{x_i} and $F_{\overline{x}_i}$ are implemented such that they are robust testable, then all path delay faults in F are robustly testable [89]. On the other hand, if F_{x_i} and $F_{\overline{x}_i}$ are implemented such

CHAPTER 9

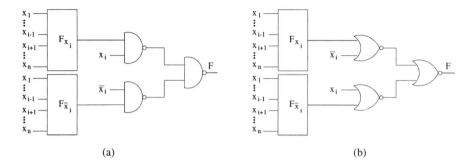

Figure 9.1. Implementations of decomposition given by (a) Equation (9.1) and (b) Equation (9.2).

that they are not robustly testable, then Shannon's expansion can be iteratively applied until the subfunctions F_{x_i} and $F_{\bar{x}_i}$ can be implemented as robustly testable circuits. Heuristics are proposed for choosing the splitting variable x_i [89].

Another approach that aims at improving the RPDF testability of multi-level circuits has been proposed by Roy et al. [130]. This approach identifies the reconvergence points in the circuit as the main cause of poor path delay fault testability. Local don't care terms are used to check if a given path is robustly testable. If a path is not testable, local transformations are used at the reconvergence points to enhance the testability. The concept of *reconvergence zone sets* is used to identify the subcircuits whose modification using local transformations can improve the RPDF testability.

DEFINITION 9.1 [130] The set of gates obtained by the intersection of the fanin cone of gate j and the fanout cone of gate i (excluding gates i and j) is called a **reconvergence zone set** for gates i and j.

The local transformations involve flattening a subcircuit into two levels, minimizing it and decomposing using Shannon's expansion theorem as earlier described.

A more common approach to obtain robust delay fault testable circuits follows the flow given in Figure 9.2 [39, 40, 74, 127]. First, a two-level robust delay testable design is synthesized and then, multi-level circuits are obtained by applying only those transformations that preserve robust delay fault testability.

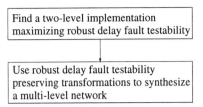

Figure 9.2. Typical synthesis process for robust path delay fault testability.

A two-level minimization process does not always lead to a completely RPDF testable circuit. For example, consider the two implementations of the same logic function F given in Figure 9.3 [39]. Both implemen-

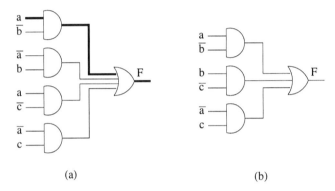

Figure 9.3. Two implementations of logic function F.

tations are prime and irredundant but only the implementation in Figure 9.3(b) is completely RPDF testable. The path shown in bold in Figure 9.3(a) is not robustly testable for a rising transition at signal a. Therefore, the two-level optimization process has to be specially tuned to obtain two-level robust path delay fault testable circuits. Necessary and sufficient conditions for synthesizing a two-level circuit such that it is completely path delay fault testable have been derived by Devadas and Keutzer [39, 40]. The same work also considers necessary and sufficient conditions for synthesizing two-level circuits that are fully transition fault testable. Next, robust delay fault testability preserving transformations are applied to obtain multi-level networks that are fully robust delay fault testable. It can be shown that algebraic factorization techniques (including cube extraction, kernel extraction and algebraic resubstitution) [18] preserve robust path delay fault testability and can be used to synthesize multi-level designs from robust testable two-level circuits. On the other hand, transformations such as eliminating a node by pushing it into its fanout nodes and Boolean factorization may diminish robust path delay fault testability of the circuit [39].

EXAMPLE 9.1 [39] Consider the circuit in Figure 9.4(a) implementing function $F = (a + c)\bar{b} + b\bar{c}$. This is a completely robust path delay fault testable circuit. If node $(a + c)$ is eliminated, the function becomes $F = a\bar{b} + c\bar{b} + b\bar{c}$ and the circuit implementing it is shown in Figure 9.4(b). In this circuit, path $\{\uparrow, \bar{b}df\}$ is not robustly testable. Therefore, node elimination does not always preserve robust path delay fault testability.

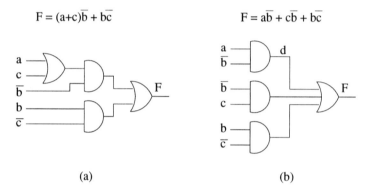

Figure 9.4. Node elimination does not preserve robust path delay fault testability.

Local transformations that preserve or improve RPDF testability of multi-level designs while optimizing the design area have been studied [59, 60]. Multi-level transition fault testable designs can be obtained from two-level testable designs using constrained algebraic factorization [40].

For some functions it is impossible to find a two-level implementation that satisfies the necessary and sufficient conditions for complete robust path delay or transition fault testability. Therefore, the initial two-level circuit might not be fully robust testable which, in turn, means that the multi-level circuit obtained using algebraic factorization is not fully robust delay testable either. Jha et al. [74] propose several synthesis rules including transformation from two-level to three- or four-level implementations to eliminate undetectable path delay faults. Algebraic factorization can then be used to obtain multi-level fully robust testable circuits.

9.1.2 Standard scan sequential circuits

Synthesizing robust transition fault testable designs under a standard scan design methodology has been considered by Cheng et al. [33].

DEFINITION 9.2 A **one-hot coding** of a finite state machine with n states, implies a code of length n such that the code for each state has a 1 in a unique identifying position and 0's in all other positions.

EXAMPLE 9.2 Consider the state transition graph of a finite state machine given in Figure 9.5(a). A one-hot encoding is shown in Figure 9.5(b). After the one-hot encoding, the unused state codes are assigned don't care conditions during two-level logic minimization. The don't care set is also shown in Figure 9.5(b). In a two-level implementation there is no sharing between the logic of the next state lines in a one-hot coded and minimized finite state machine. However, a cube (AND gate) may feed both PO and NS lines.

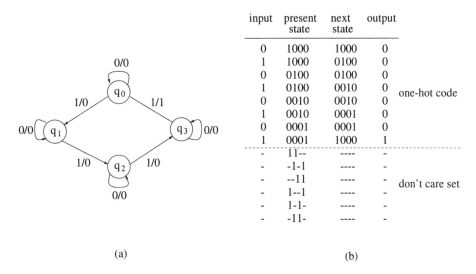

Figure 9.5. One-hot coding of a finite state machine.

Cheng et al. [33] show that a one-hot coded and optimized finite state machine (FSM) whose state transition graph (STG) satisfies a certain property is guaranteed to be fully robust transition fault testable under standard scan. Any state transition graph that does not satisfy the above property can be transformed into a state transition graph that satisfies the property by an appropriate addition of edges and states.

Requiring the state transition graph to satisfy the above property is a sufficiency condition, and not a necessary one for a fully transition fault testable machine. It is assumed that, under the standard scan design methodology, the state part s_2 of the second vector v_2 of the vector pair $\langle v_1, v_2 \rangle = \langle i_1 + s_1, i_2 + s_2 \rangle$ is generated using either scan shifting or functional justification (see Chapter 2).

Given a two-level circuit and an AND gate g in that circuit, a $0 \to 1$ transition fault on g can be detected at a primary output or next state (NS) line by applying vector pair $\langle v_1, v_2 \rangle$, where vector v_1 is in the OFF-set of the PO or NS line, and vector v_2 is an irredundancy test for the cube corresponding to g for the PO or NS line. An irredundancy test implies that all AND gates that feed the PO or NS line are held at 0, except for g which is at 1. Similarly, applying vector pair $\langle v_2, v_1 \rangle$ will detect $1 \to 0$ transition fault on g. The above test, $\langle v_1, v_2 \rangle$, may have hazards and therefore, it may not be a robust test for the transition fault on g. However, such tests may be allowed in order to avoid the use of multi-valued logic in test generation.

Testing any AND gate in a single-output two-level cover will simultaneously test the OR gate as well.

Let C represent a one-hot coded and two-level minimized cover for the state transition graph of a given finite state machine. Consider an AND gate g in C. There are four possible cases for testing the AND gate g:

(1) test g for a $1 \to 0$ transition fault to a PO line o,

(2) test g for a $1 \to 0$ transition fault to an NS line n,

(3) test g for a $0 \to 1$ transition fault to a PO line o,

(4) test g for a $0 \to 1$ transition fault to an NS line n.

The requirements on $\langle i_1 + s_1, i_2 + s_2 \rangle$ for testing a particular AND gate g for each of the four cases above are as follows:

(1) An irredundancy test $i_1 + s_1$ for g at PO line o, should produce state s_2, which satisfies the property that there exists i_2 such that the value on primary output o on the application of vector $i_2 + s_2$ is 0.

If state s_2 does not satisfy this property, the original STG G can be modified such that each state $q \in G$ produces a value 0 for each output o on at least one input minterm i_2. Note that it is not required that both 0 and 1 values on o are produced by each state q, nor that the minterm producing the 0 value is the same for different states. Also, there may be several irredundancy tests for gate g and only one irredundancy test is required to produce s_2 that satisfies the above property. Making each state in STG satisfy the property is an easy sufficiency condition. In order to relax this condition, it is necessary to *predict* what the irredundancy tests in the one-hot coded and minimized cover are and relate them to input minterms and states which is a difficult task.

(2) An irredundancy test for g at the NS line n, $i_1 + s_1$, will result in the machine going to state s_n, which is the state corresponding to the NS line n. This is because in a one-hot coding there is a unique coding for each state line and if the irredundancy test produces a 1 at the NS line n, it means that the FSM has moved to state s_n. Next, there must exist i_2 such that vector $i_2 + s_n$ takes the machine to a state $s_3 \neq s_n$. If $s_3 \neq s_n$, then the NS line n will have the value 0. This requirement is easy to satisfy. For each state $q \in G$, it is required that there is at least one input minterm i_2 that moves the machine to some state other than q. In other words, there should be no state with only loop edges. A strongly connected machine [67], in which every state can be reached from every other state satisfies the above property. A STG that is not strongly connected can easily be modified to make it strongly connected.

(3) Consider an irredundancy test for g at a PO line o, $i_2 + s_2$. Then state s_1 must be produced by some input vector i_1 from some tate s_1 such that the value of o is 0. In other words, for each state q in the STG, there should exist at least one incoming edge which produces a value 0 for each PO o. Again, a STG can be easily modified to satisfy this sufficient condition.

(4) Consider an irredundancy test $i_2 + s_2$ for g corresponding to a NS line n. This vector results in the machine going into state s^n corresponding to the NS line n. If $s^n \neq s_2$ then it is possible to reach s_2 with the value of the NS line n at 0, and the application of i_2.

As mentioned earlier, it is difficult to predict what the irredundancy tests of each cube in on-hot coded and minimized cover are going to be. In this case, the irredundancy tests for a cube and an NS line n should *not* correspond to s^n. If a loop edge of s^n becomes an irredundancy test for NS line n, then the above condition is violated.

An easy way to ensure that this does not happen is to split the states in the machine such that all loop edges are removed.

The following example illustrates the application of the above transformations to obtain a fully transition fault testable machine.

EXAMPLE 9.3 Consider the STG shown in Figure 9.6(a). It satisfies the above conditions 1 and 2. State s_1 does not satisfy condition 3, and an edge from state s_2 fanning out to s_1 is added to satisfy the property. Finally, state s_3 is split into states s_{3a} and s_{3b} to satisfy the condition 4. One-hot encoding and minimization of the

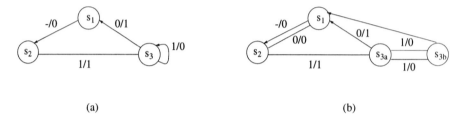

(a)   (b)

Figure 9.6. Transformations to achieve testability under one-hot coding.

STG shown in Figure 9.6(b) will result in a fully transition fault testable machine.

One-hot codes may require too many bits and may be area inefficient for large finite state machines. The above results can be further extended to arbitrary length encodings which, together with a good state assignment algorithm, can lead to highly robust transition fault testable designs [33].

9.2 SYNTHESIS FOR VALIDATABLE NON-ROBUST TESTABLE AND DELAY-VERIFIABLE CIRCUITS

In most designs robustly testable path delay faults represent only a small fraction of all path delay faults. Therefore, synthesis for robust testability often results in large area and performance overheads or a large number of additional primary inputs. The necessary and sufficient conditions for validatable non-robust (VNR) testability (see Chapter 5) of designs are fewer stringent than those for robust path delay fault testability [41]. They impose fewer constraints on the synthesis process and can result in more area efficient circuits. However, since the necessary and sufficient conditions to achieve VNR path delay fault testability require that each output function in the two-level representation must be prime and irredundant [41], the amount of logic that can be shared is reduced. Multi-level circuits can be obtained by applying algebraic factorization to the initial two-level fully VNR testable circuit.

Another way to reduce the area overhead is the synthesis for delay-verifiability considered by Ke and Menon [79]. A circuit is said to be *delay-verifiable* if after applying tests for robust and validatable non-robust path delay faults and tests for primitive faults of cardinality higher than 1, the circuit performance can be guaranteed. Non-robustly testable paths that cannot be validated are not allowed in the design. Obtaining delay-verifiable multi-level circuits requires identification of all primitive faults of cardinality higher than 1. As discussed in Chapter 7, for large designs, this is a complicated problem. However, identifying primitive faults in two-level circuits can be easily done. Necessary and sufficient conditions for primitive faults in two-level designs have been presented in [79]. Identification of primitive faults in two-level representations can be done by checking the primality of the literal set from which the paths start and the irredundancy of the corresponding cubes. The synthesis procedure for delay-verifiable circuits proposed by Ke and Menon follows a flow similar to that in Figure 9.2. First, a delay-verifiable two-level circuit is found and then algebraic factorization is used to obtain a delay-verifiable multi-level circuit. As it was the case with other similar approaches, the need to obtain a two-level delay-verifiable representation limits the applicability of this synthesis method. This is because many functions cannot be represented in a two-level form of a reasonable size.

9.3 SUMMARY

Fully delay fault testable designs usually come at the expense of area and/or performance. Most of the synthesis techniques for delay fault testability try to achieve designs in which all faults are testable. The cost, however, might be lower if instead of synthesizing circuits that are completely delay fault testable, circuits are synthesized such that they also contain some faults that can affect the performance only if some of the testable faults are present. Techniques for identifying path delay faults that cannot independently determine the performance of the design have been discussed in Chapter 5. Also, most of the proposed synthesis techniques for delay faults rely on achieving a fully delay testable two-level design prior to synthesizing a multi-level delay testable circuit.

However, for many practical designs, this strategy is not feasible because representing them as two-level circuits cannot be done using reasonable computing resources.

10 CONCLUSIONS AND FUTURE WORK

Delay testing is becoming an increasingly important part of VLSI design testing process. Continuously increasing circuit operating frequencies result in designs in which performance specifications can be violated by very small defects. Studies show that high stuck-at fault coverage is not sufficient to guarantee detection of these timing failures. The use of traditional fault models and testing strategies becomes even more inadequate as the current design trends move towards deeper submicron designs. The deep submicron process introduces new failure modes and a new set of design and test problems. Process variations are now more likely to cause marginal violations of the performance specifications. Continuous shrinking of device feature size, increased number of interconnect layers and gate density, increased current density and higher voltage drop along the power nets give rise to the "noise faults" such as: distributed delay defects, power bus noise, ground bounce, substrate noise and crosstalk. Analysis shows that excessive noise most of the time leads to delay faults [19]. For example, studies have shown that the increased coupling effects produce interference between signals and may increase or decrease signal delays [19]. As an illustration, consider the simulation model shown in Figure 10.1 [19]. Capacitive coupling be-

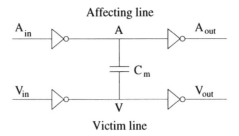

Figure 10.1. Simulation model for crosstalk.

tween the affecting line (A) and the victim line (V) can result in crosstalk effects such as overshoot, undershoot, glitches, speedup and slowdown. Simulation results show that when A_{in} and V_{in} have transitions in the same direction at the same time, a speedup occurs on both lines. Also, when A_{in} and V_{in} have opposite transitions at the same time, a slowdown occurs on both lines. Speedup and slowdown are measured with respect to the nominal delay that corresponds to the situation in which one of the lines is static and the other has a transition. Both slowdown and speedup can cause errors. Signal slowdown can cause delay faults if a transition is propagated along paths with small slacks while signal speedup can cause race conditions if transitions are propagated along short paths.

A simulation example by Breuer et al. [19] shows a 50% decrease and 86% increase in the signal delay (compared to the signal nominal delay) as a result of crosstalk induced speedup and slowdown. Figure 10.2 [19] shows the effect of the time skew z between the input signals on the crosstalk speedup/slowdown on the affecting line A. Step inputs are assumed at both inputs. Figure 10.2(a) illustrates the case when the transition on signal A_{in} occurs before the transition on signal V_{in}, i.e., $z < 0$. It can be seen that as $|z|$ increases, the slowdown increases while the speedup decreases. Figure 10.2(a) illustrates the case when $z > 0$, i.e., the transition on signal A_{in} occurs after the transition on signal V_{in}. In this case, the increase in z results in a decrease of the slowdown and an increase of the speedup. Furthermore, the study shows that crosstalk induced speedup/slowdown depends on the signal fall/rise delays. Faster input transitions result in a greater crosstalk induced speedup/slowdown relative to the nominal delay time. Also, experiments show that process variations intensify noise and crosstalk speedup/slowdown effects become even more pronounced. Other types of "noise faults" also affect the signal delays.

CONCLUSIONS AND FUTURE WORK 171

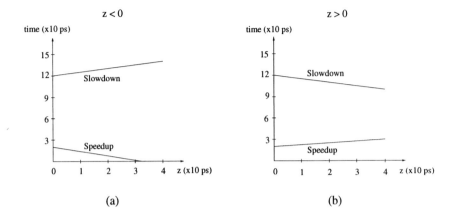

Figure 10.2. Effect of time skew z between the input signals on crosstalk speedup/slowdown on affecting line for (a) $z < 0$ and (b) $z > 0$.

Delay testing models and testing techniques and tools currently in use cannot satisfactorily address the new failure modes in deep submicron designs. They suffer from serious deficiencies:

- Testing is usually limited to transition faults and robust and non-robust path delay faults. Transition faults can detect large delay defects but they are not suitable for detecting small distributed defects resulting from process variations. Also, in many designs most of the path delay faults cannot be sensitized under robust and non-robust propagation conditions but can degrade the circuit performance and tests for them need to be also included.

- No practical solutions exist for low fault coverage of delay tests. Most of the proposed design and synthesis methods for delay fault testability techniques result in an high area/performance overhead and/or large number of additional input pins.

- Selection of critical path delay faults is based on very simple and inaccurate delay models. Since testing all path delay faults is infeasible for most designs, selecting truly timing critical faults for test becomes crucial for high product quality levels.

- Most of the available delay test solutions are only applicable to full-scan circuits. Extending the existing delay test techniques for non-scan sequential designs is a complex problem.

172   CHAPTER 10

- The cost of testers is directly proportional to their speed and often delay testing has to be performed using testers that are slower than the speed of the new designs.

- Delay faults caused by crosstalk and power supply noise are completely ignored.

Some of the above issues have been addressed by recent research and selected results have been described in this book. Most of the existing techniques are based on simplified, logic-level delay fault models and cannot be directly used to model and test timing defects in high speed designs based on deep submicron technologies. Therefore, new delay testing research directions should attempt to close the gap between the logic-level delay fault model and the physical defects. The existing delay testing research results can then be adapted using new, more accurate models. Possible future delay fault testing research topics include:

- Developing new path delay fault selection and test generation methods. These methods should take into account the performance degradation caused by random delay defects as well as by power net voltage drops, ground bounce, process variation and interconnect crosstalk.

- Developing new, practical solutions for testing fast chips on slow testers. Experimental results show that the existing techniques result in unacceptably low fault coverages. At the current rate of design performance increase and the high cost of fast testers, the gap between the speed of the new designs and that of the testers is not likely to disappear.

- Extensions of delay testing techniques for handing partial scan and non-scan circuits need to be addressed. The variable clock [26] and rated-clock [15] delay test procedures for non-scan circuits are yet to be put to real practice.

REFERENCES

[1] M. Abramovici, M. A. Breuer, and A. D. Friedman. *Digital Systems Testing and Testable Design*. Computer Science Press, New York, 1990.

[2] L. Ackner and M. R. Barber. Frequency Enhancement of Digital VLSI Test Systems. *Proceedings of IEEE International Test Conference*, pages 444–451, October 1990.

[3] P. Agrawal, V. D. Agrawal, and S. C. Seth. DynaTAPP: Dynamic Timing Analysis with Partial Path Activation in Sequential Circuits. *Proceedings of European Design Automation Conference*, pages 138–141, September 1992.

[4] P. Agrawal, V. D. Agrawal, and S. C. Seth. Generating Tests for Delay Faults in Nonscan Circuits. *IEEE Design & Test of Computers*, 10(1):20–28, March 1993.

[5] P. Agrawal and W. J. Dally. A Hardware Logic Simulation System. *IEEE Transactions on Computer-Aided Design of Integrated Circuits and Systems*, 9(1):19–29, January 1990.

[6] V. D. Agrawal and T. J. Chakraborty. High-Performance Circuit Testing with Slow-Speed Testers. *Proceedings of IEEE International Test Conference*, pages 302–310, October 1995.

[7] V. D. Agrawal and S. T. Chakradhar. Combinational ATPG Theorems for Identifying Untestable Faults in Sequential Circuits. *IEEE Transactions on Computer-Aided Design of Integrated Circuits and Systems*, 14(9):1115–1127, September 1995.

[8] V. D. Agrawal, C.-J. Lin, P. W. Rutkowski, S. Wu, and Y. Zorian. Built-In Self-Test for Digital Integrated Circuits. *AT&T Technical Journal*, 73:30–39, March 1994.

[9] K. A. Bartlett et al. Multilevel Logic Minimizing Using Implicit Don't Cares. *IEEE Transactions on Computer-Aided Design of Integrated Circuits and Systems*, 7(6):723–740, June 1988.

[10] S. Barton. Characterization of High-Speed (Above 500 MHz) Devices using Advanced ATE - Technique, Results and Device Problems. *Proceedings of IEEE International Test Conference*, pages 860–868, October 1989.

[11] R. Bencivenga, T. J. Chakraborty, and S. Davidson. The Architecture of the Gentest Sequential Circuit Test Generator. *Proceedings of Custom Integrated Circuits Conference*, pages 17.1.1–17.1.4, May 1991.

[12] S. Bose. Testing for Path Delay Faults in Synchronous Digital Circuits. *PhD Dissertation*, December 1995. Carnegie Mellon University, Pittsburgh, PA.

[13] S. Bose, P. Agrawal, and V. D. Agrawal. Delay Fault Testability Evaluation Through Timing Simulation. *Proceedings of 3rd Great Lakes Symposium on VLSI*, pages 18–21, March 1993.

[14] S. Bose, P. Agrawal, and V. D. Agrawal. Path Delay Fault Simulation of Sequential Circuits. *IEEE Transactions on VLSI Systems*, 1(4):453–461, December 1993.

[15] S. Bose, P. Agrawal, and V. D. Agrawal. A Rated-Clock Test Method for Path Delay Faults. *IEEE Transactions on VLSI Systems*, 6(6):323–331, June 1998.

[16] S. Bose and V. D. Agrawal. Sequential Logic Path Delay Test Generation by Symbolic Analysis. *Proceedings of 4th Asian Test Symposium*, pages 353–359, November 1995.

[17] S. Bose, V. D. Agrawal, and T. G. Szymanski. Algorithms for Switch Level Delay Fault Simulation. *Proceedings of IEEE International Test Conference*, pages 982–991, November 1997.

[18] R. Brayton, R. Rudell, A. Sangiovanni-Vincentelli, and A. Wang. MIS: A Multiple-Level Logic Optimization System. *IEEE Transactions on Computer-Aided Design of Integrated Circuits and Systems*, CAD-6(11):1062–1081, November 1987.

[19] M. A. Breuer, C. Gleason, and S. Gupta. New Validation and Test Problems for High Performance Deep Sub-micron VLSI Circuits. *Tutorial Notes, 15th IEEE VLSI Test Symposium*, April 1997.

[20] M. A. Breuer and N. K. Nanda. Simplified Delay Testing for LSI Circuit Faults. *United States Patent Number 4,672,307*, June 1987.

[21] O. Bula, J. Moser, J. Trinko, M. Weissman, and Woytowich. Gross Delay Defect Evaluation for a CMOS Logic Design System Product. *IBM Journal of Research and Development*, 34(2-3):325–338, March–May 1990.

[22] J. L. Carter, V. S. Iyengar, and B. K. Rosen. Efficient Test Coverage Determination for Delay Faults. *Proceedings of IEEE International Test Conference*, pages 418–427, September 1987.

[23] T. J. Chakraborty, V. D. Agrawal, and M. L. Bushnell. Path Delay Fault Simulation Algorithms for Sequential Circuits. *Proceedings of 1st Asian Test Symposium*, pages 52–56, November 1992.

[24] T. J. Chakraborty, V. D. Agrawal, and M. L. Bushnell. Delay Fault Models and Test Generation of Random Logic Sequential Ciruits. *Proceedings of 30th Design Automation Conference*, pages 453–457, June 1993.

[25] T. J. Chakraborty, V. D. Agrawal, and M. L. Bushnell. Design for Testability for Path Delay Faults in Sequential Circuits. *Proceedings of 30th Design Automation Conference*, pages 453–457, June 1993.

[26] T. J. Chakraborty, V. D. Agrawal, and M. L. Bushnell. On Variable Clock Methods for Path Delay Testing of Sequential Circuits.

IEEE Transactions on Computer-Aided Design of Integrated Circuits and Systems, 16(11):1237–1249, November 1997.

[27] S.-C. Chang, M. Marek-Sadowska, and K.-T. Cheng. Perturb and Simplify: Multilevel Boolean Network Optimizer. *IEEE Transactions on Computer-Aided Design of Integrated Circuits and Systems*, 15(12):1494–1504, December 1996.

[28] L.-C. Chen, S. K. Gupta, and M. A. Breuer. High Quality Robust Tests for Path Delay Faults. *Proceedings of 15th IEEE VLSI Test Symposium*, pages 88–93, May 1997.

[29] K.-T. Cheng. Redundancy Removal for Sequential Circuits Without Reset States. *IEEE Transactions on Computer-Aided Design of Integrated Circuits and Systems*, 12(1):13–24, January 1993.

[30] K.-T. Cheng. Transition Fault Testing for Sequential Circuits. *IEEE Transactions on Computer-Aided Design of Integrated Circuits and Systems*, 12(12):1971–1983, December 1993.

[31] K.-T. Cheng and V. D. Agrawal. A Partial Scan Method for Sequential Circuits with Feedback. *IEEE Transactions on Computers*, 39(4):544–548, April 1990.

[32] K.-T. Cheng and H.-C. Chen. Classification and Identification of Nonrobust Untestable Path Delay Faults. *IEEE Transactions on Computer-Aided Design of Integrated Circuits and Systems*, 15(8):845–853, August 1996.

[33] K.-T. Cheng, S. Devadas, and K. Keutzer. Delay-Fault Test Generation and Synthesis for Testability Under a Standard Scan Design Methodology. *IEEE Transactions on Computer-Aided Design of Integrated Circuits and Systems*, 12(8):1217–1231, August 1993.

[34] K.-T. Cheng, A. Krstić, and H.-C. Chen. Generation of High Quality Tests for Robustly Untestable Path Delay Faults. *IEEE Transactions on Computers*, 45(12):1379–1392, December 1996.

[35] K.-T. Cheng and C.-J. Lin. Timing-Driven Test Point Insertion for Full-Scan and Partial-Scan BIST. *Proceedings of IEEE International Test Conference*, pages 506–514, October 1995.

[36] G. De Micheli. *Synthesis and Optimization of Digital Circuits*. McGraw-Hill, New York, 1994.

[37] B. I. Dervisoglu and G. E. Strong. Design for Testability: Using Scanpath Techniques for Path-Delay Test and Measurement. *Proceedings of IEEE International Test Conference*, pages 365–374, October 1991.

[38] S. Devadas. Delay Test Generation for Synchronous Sequential Circuits. *Proceedings of IEEE International Test Conference*, pages 144–152, September 1989.

[39] S. Devadas and K. Keutzer. Synthesis of Robust Delay-Fault Testable Circuits: Practice. *IEEE Transactions on Computer-Aided Design of Integrated Circuits and Systems*, 11(3):277–300, March 1992.

[40] S. Devadas and K. Keutzer. Synthesis of Robust Delay-Fault Testable Circuits: Theory. *IEEE Transactions on Computer-Aided Design of Integrated Circuits and Systems*, 11(1):87–101, January 1992.

[41] S. Devadas and K. Keutzer. Validatable Nonrobust Delay-Fault Testable Circuits Via Logic Synthesis. *IEEE Transactions on Computer-Aided Design of Integrated Circuits and Systems*, 11(12):1559–1573, December 1992.

[42] S. Devadas, K. Keutzer, and S. Malik. Computation of Floating Mode Delay in Combinational Circuits: Theory and Algorithms. *IEEE Transactions on Computer-Aided Design of Integrated Circuits and Systems*, 12(12):1913–1923, December 1993.

[43] D. Dumas, P. Girard, C. Landrault, and S. Pravossoudovitch. An Implicit Delay Fault Simulation Method with Approximate Detection Threshold Calculation. *Proceedings of IEEE International Test Conference*, pages 705–713, October 1993.

[44] L. A. Entrena and K.-T. Cheng. Combinational and Sequential Logic Optimization by Redundancy Addition and Removal. *IEEE Transactions on Computer-Aided Design of Integrated Circuits and Systems*, 14(7):909–916, July 1995.

[45] F. Fink, K. Fuchs, and M. H. Schulz. An Efficcient Parallel Pattern Gate Delay Fault Simulator with Accelerated Detected Fault Size Detremination Capabilities. *Proceedings of European Test Conference*, pages 171–180, April 1991.

[46] F. Fink, K. Fuchs, and M. H. Schulz. Robust and Nonrobust Path Delay Fault Simulation by Parallel Processing of Patterns. *IEEE Transactions on Computers*, 41(12):1527–1536, December 1992.

[47] P. Franco, W. D. Farwell, R. L. Stokes, and E. J. McCluskey. An Experimental Chip to Evaluate Test Techniques: Chip and Experiment Design. *Proceedings of IEEE International Test Conference*, pages 653–662, October 1995.

[48] P. Franco, S. Ma, J. Chang, Y.-C. Chu, S. Wattal, E. J. McCluskey, R. L. Stokes, and W. D. Farwell. Analysis and Detection of Timing Failures in An Experimental Test Chip. *Proceedings of IEEE International Test Conference*, pages 691–700, October 1996.

[49] P. Franco and E. J. McCluskey. Three-Pattern Tests for Delay Faults. *Proceedings of 12th IEEE VLSI Test Symposium*, pages 452–456, April 1994.

[50] K. Fuchs, F. Fink, and M. H. Schulz. DYNAMITE: An Efficient Automatic Test Pattern Generation System for Path Delay Faults. *IEEE Transactions on Computer-Aided Design of Integrated Circuits and Systems*, CAD–10:1323–1335, October 1991.

[51] J. A. Gasbarro and M. A. Horowitz. Techniques for Characterizing DRAMS with a 500 MHz Interface. *Proceedings of IEEE International Test Conference*, pages 516–525, October 1994.

[52] M. A. Gharaybeh, V. D. Agrawal, and M. L. Bushnell. False-Path Removal using Delay Fault Simulation. *Proceedings of 7th Asian Test Symposium*, December 1998.

[53] M. A. Gharaybeh, M. L. Bushnell, and V. D. Agrawal. Classification and Modeling of Path Delay Faults and Test generation Using Single Stuck-Fault Tests. *Proceedings of IEEE International Test Conference*, pages 139–148, October 1995.

[54] M. A. Gharaybeh, M. L. Bushnell, and V. D. Agrawal. An Exact Non-Enumerative Fault Simulator for Path-Delay Faults. *Proceedings of IEEE International Test Conference*, pages 276–285, October 1996.

[55] M. A. Gharaybeh, M. L. Bushnell, and V. D. Agrawal. Classification and Modeling of Path Delay Faults and Test generation Using

Single Stuck-Fault Tests. *Journal of Electronic Testing: Theory and Applications*, 11(1):55–67, August 1997.

[56] M. A. Gharaybeh, M. L. Bushnell, and V. D. Agrawal. The Path-Status Graph with Application to Delay Fault Simulation. *IEEE Transactions on Computer-Aided Design of Integrated Circuits and Systems*, 17(4), April 1998.

[57] P. Goel. An Implicit Enumeration Algorithm to Generate Tests for Combinational Logic Circuits. *IEEE Transactions on Computers*, C-30(3):215–222, March 1981.

[58] H. Hao and E. J. McCluskey. Very Low Voltage Testing for Weak CMOS Logic ICs. *Proceedings of IEEE International Test Conference*, pages 275–284, October 1993.

[59] H. Hengster, R. Drechsler, and B. Becker. Testability Properties of Local Circuit Transformations with Respect to the Robust Path-Delay-Fault Model. *Proceedings of 7th International Conference on VLSI Design*, pages 123–126, January 1994.

[60] H. Hengster, R. Drechsler, and B. Becker. On the Application of Local Circuits Transformations with Special Emphasis on Path Delay Fault Testability. *Proceedings of 13th IEEE VLSI Test Symposium*, pages 378–392, April 1995.

[61] K. Heragu, V. D. Agrawal, M. L. Bushnell, and J. H. Patel. Improving a Nonenumerative Method to Estimate Path Delay Fault Coverage. *IEEE Transactions on Computer-Aided Design of Integrated Circuits and Systems*, 16(7):759–762, July 1997.

[62] K. Heragu, J. H. Patel, and V. D. Agrawal. An Efficient Path Delay Fault Coverage Estimator. *Proceedings of 31st Design Automation Conference*, pages 516–521, June 1994.

[63] K. Heragu, J. H. Patel, and V. D. Agrawal. Improving Accuracy in Path Delay Fault Coverage Estimation. *Proceedings of 9th International Conference on VLSI Design*, pages 422–425, January 1996.

[64] K. Heragu, J. H. Patel, and V. D. Agrawal. Segment Delay Faults: A New Fault Model. *Proceedings of 14th IEEE VLSI Test Symposium*, pages 32–39, May 1996.

[65] K. Heragu, J. H. Patel, and V. D. Agrawal. SIGMA: A Simulator for Segment Delay Faults. *Proceedings of IEEE/ACM International Conference on Computer-Aided Design*, pages 502–508, November 1996.

[66] K. Heragu, J. H. Patel, and V. D. Agrawal. Fast Identification of Untestable Delay Faults Using Implications. *Proceedings of IEEE/ACM International Conference on Computer-Aided Design*, pages 642–647, November 1997.

[67] F. J. Hill and G. R. Peterson. *Introduction to Switching Theory and Logical Design*. Wiley, New York, 1981.

[68] Y.-H. Hsu and S. K. Gupta. A Simulator for At-Speed Robust Testing of Path Delay Faults in Combinational Circuits. *IEEE Transactions on Computers*, 45(11):1312–1318, November 1996.

[69] V. S. Iyengar, B. K. Rosen, and I. Spillinger. Delay Test Generation 1 – Concepts and Coverage Metrics. *Proceedings of IEEE International Test Conference*, pages 857–866, September 1988.

[70] V. S. Iyengar, B. K. Rosen, and I. Spillinger. Delay Test Generation 2 – Algebra and Algorithms. *Proceedings of IEEE International Test Conference*, pages 867–876, September 1988.

[71] V. S. Iyengar, B. K. Rosen, and J. A. Waicukauski. On Computing the Sizes of Detected Delay Faults. *IEEE Transactions on Computer-Aided Design of Integrated Circuits and Systems*, 9(3):299–312, March 1990.

[72] V. S. Iyengar and G. Vijayan. Optimized Test Application Timing for AC Test. *IEEE Transactions on Computer-Aided Design of Integrated Circuits and Systems*, 11(11):1439–1449, November 1992.

[73] M. A. Iyer and M. Abramovici. Low-Cost Redundancy Identification for Combinational Circuits. *Proceedings of 7th International Conference on VLSI Design*, pages 315–318, January 1994.

[74] N. K. Jha, I. Pomeranz, S. M. Reddy, and R. J. Miller. Synthesis of Multi-Level Combinational Circuits for Complete Robust Path Delay Fault Testability. *Proceedings of IEEE Workshop on Fault-Tolerant Parallel and Distributed Systems*, pages 280–287, July 1992.

[75] D. Kagaris, S. Tragoudas, and D. Karayiannis. Improved Non-Enumerative Path-Delay Fault-Coverage Estimation Based on Optimal Polynomial-Time Algorithms. *IEEE Transactions on Computer-Aided Design of Integrated Circuits and Systems*, 16(3):309–315, March 1997.

[76] D. Kagaris, S. Tragoudas, and D. Karayiannis. Nonenumerative Path Delay Fault Coverage Estimation with Optimal Algorithms. *Proceedings of IEEE International Conference on Computer Design*, pages 366–371, October 1997.

[77] B. Kapoor. An Efficient Method for Computing Exact Path Delay Fault Coverage. *Proceedings of European Design and Test Conference*, pages 516–520, March 1995.

[78] W. Ke and P. R. Menon. Delay-Verifiability of Combinational Circuits Based on Primitive Faults. *Proceedings of IEEE International Conference on Computer Design*, pages 86–90, October 1994.

[79] W. Ke and P. R. Menon. Synthesis of Delay-Verifiable Combinational Circuits. *IEEE Transactions on Computers*, 44(2):213–222, February 1995.

[80] D. C. Keezer. Multiplexing Test System Channels for Data Rates Above 1Gb/s. *Proceedings of IEEE International Test Conference*, pages 790–797, October 1990.

[81] T. Kirkland and M. R. Mercer. A Topological Search Algorithm for ATPG. *Proceedings of 24th Design Automation Conference*, pages 502–508, June 1987.

[82] Z. Kohavi. *Switching and Finite Automata Theory*. McGraw-Hill, New York, 1978.

[83] A. Krstić, S. T. Chakradhar, and K.-T. Cheng. Testable Path Delay Fault Cover for Sequential Circuits. *Proceedings of European Design Automation Conference*, pages 220–226, September 1996.

[84] A. Krstić, S. T. Chakradhar, and K.-T. Cheng. Design for Primitive Delay Fault Testability. *Proceedings of IEEE International Test Conference*, pages 436–445, November 1997.

[85] A. Krstić and K.-T. Cheng. Resynthesis of Combinational Circuits for Path Count Reduction and for Path Delay Fault Testability.

Proceedings of European Design and Test Conference, pages 486–490, March 1996.

[86] A. Krstić and K.-T. Cheng. Resynthesis of Combinational Circuits for Path Count Reduction and for Path Delay Fault Testability. *Journal of Electronic Testing: Theory and Applications*, pages 43–54, August 1997.

[87] A. Krstić, K.-T. Cheng, and S. T. Chakradhar. Identification and Test Generation for Primitive Faults. *Proceedings of IEEE International Test Conference*, pages 423–432, October 1996.

[88] A. Krstić, K.-T. Cheng, and S. T. Chakradhar. Testing High Speed VLSI Devices Using Slower Testers. *Technical Report No. 98-18*, 1998. University of California, Santa Barbara.

[89] S. Kundu, S. M. Reddy, and N. K. Jha. Design of Robustly Testable Combinational Logic Circuits. *IEEE Transactions on Computer-Aided Design of Integrated Circuits and Systems*, 10(8):1036–1048, August 1991.

[90] W. Kunz and D. K. Pradhan. Recursive Learning: A New Implication Technique for Efficient Solutions to CAD Problems - Test, Verification and Optimization. *IEEE Transactions on Computer-Aided Design of Integrated Circuits and Systems*, 13(9):1143–1158, September 1994.

[91] W. Kunz and D. Stoffel. *Reasoning in Boolean Networks: Logic Synthesis and Verification Using Testing Techniques*. Kluwer Academic Publishers, Boston, 1997.

[92] W. K. Lam, A. Saldanha, R. K. Brayton, and A. L. Sangiovanni-Vicentelli. Delay Fault Coverage, Test Set Size, and Performance Trade-Offs. *IEEE Transactions on Computer-Aided Design of Integrated Circuits and Systems*, 14(1):32–44, January 1995.

[93] J. P. Lesser and J. J. Shedletsky. An Experimental Delay Test Generator for LSI Logic. *IEEE Transactions on Computers*, C-29(3):235–248, March 1980.

[94] Y. Levendel and P. R. Menon. Transition Faults in Combinational Circuits: Input Transition Test Generation and Fault Simulation. *Proceedings of International Fault Tolerant Computing Symposium*, pages 278–283, July 1986.

[95] W.-N. Li, S. M. Reddy, and S. K. Sahni. On Path Selection in Combinational Logic Circuits. *IEEE Transactions on Computer-Aided Design of Integrated Circuits and Systems*, 8(1):56–63, January 1989.

[96] Z. Li, Y. Min, and R. K. Brayton. Efficient Identification of Non-Robustly Untestable Path Delay Faults. *Proceedings of IEEE International Test Conference*, pages 992–997, November 1997.

[97] H.-C. Liang, C. L. Lee, and J. E. Chen. Identifying Untestable Faults in Sequential Circuits. *IEEE Design & Test of Computers*, 12(3):14–23, Fall 1995.

[98] C. J. Lin and S. M. Reddy. On Delay Fault Testing in Logic Circuits. *IEEE Transactions on Computer-Aided Design of Integrated Circuits and Systems*, CAD-6(5):694–703, September 1987.

[99] S. C. Ma, P. Franco, and E. J. McCluskey. An Experimental Chip to Evaluate Test Techniques: Experiment Results. *Proceedings of IEEE International Test Conference*, pages 663–672, October 1995.

[100] A. K. Majhi and V. D. Agrawal. Tutorial: Delay Fault Models and Coverage. *Proceedings of 11th International Conference on VLSI Design*, pages 364–369, January 1998.

[101] A. K. Majhi, J. Jacob, L. M. Patnaik, and V. D. Agrawal. On Test Coverage of Path Delay Faults. *Proceedings of 9th International Conference on VLSI Design*, pages 418–421, January 1996.

[102] S. Majumder, V. D. Agrawal, and M. L. Bushnell. On Delay-Untestable Paths and Stuck-Fault Redundancy. *Proceedings of 16th IEEE VLSI Test Symposium*, pages 194–199, April 1998.

[103] S. Majumder, V. D. Agrawal, and M. L. Bushnell. Path Delay Testing: Variable-Clock Versus Rated-Clock. *Proceedings of 11th International Conference on VLSI Design*, pages 470–475, January 1998.

[104] Y. K. Malaiya and R. Narayanaswamy. Modeling and Testing for Timing Faults in Synchoronous Sequential Circuits. *Design & Test of Computers*, 1(4):62–74, November 1984.

[105] W.-W. Mao and M. D. Ciletti. A Simplified Six-Waveform Type Method for Delay Fault Testing. *Proceedings of 26th Design Automation Conference*, pages 730–733, June 1989.

[106] W.-W. Mao and M. D. Ciletti. A Variable Observation Time Method for Testing Delay Faults. *Proceedings of 27th Design Automation Conference*, pages 728–731, June 1990.

[107] P. C. Maxwell, R. C. Aitken, V. Johansen, and I. Chiang. The Effect of Different Test Sets on Quality Level Prediction: When is 80% Better Than 90%. *Proceedings of IEEE International Test Conference*, pages 358–364, October 1991.

[108] P. C. Maxwell, R. C. Aitken, V. Johansen, and I. Chiang. The Effectiveness of IDDQ, Functional and Scan Tests: How Many Fault Coverages Do We Need? *Proceedings of IEEE International Test Conference*, pages 168–177, October 1992.

[109] P. C. Maxwell, R. C. Aitken, R. Kollitz, and A. C. Brown. IDDQ and AC Scan: The War Against Unmodelled Defects. *Proceedings of IEEE International Test Conference*, pages 250–258, October 1996.

[110] N. Mukherjee, T. J. Chakraborty, and S. Bhawmik. A Built-In Self-Test Methodology for Path-Delay Faults. *Proceedings of IEEE International Test Conference*, October 1998.

[111] T. M. Niermann, W.-T. Cheng, and J. H. Patel. PROOFS: A Fast, Memory-Efficient Sequential Circuit Fault Simulator. *IEEE Transactions on Computer-Aided Design of Integrated Circuits and Systems*, 11(2):198–207, February 1992.

[112] E. S. Park and M. R. Mercer. Robust and Nonrobut Tests for Path Delay Faults in a Combinational Circuit. *Proceedings of IEEE International Test Conference*, pages 1027–1034, September 1987.

[113] E. S. Park, B. Underwood, T. W. Williams, and M. R. Mercer. Delay Testing Quality in Timing-Optimized Designs. *Proceedings of IEEE International Test Conference*, pages 897–905, October 1991.

[114] C. G. Parodi, V. D. Agrawal, M. L. Bushnell, and S. Wu. A Non-Enumerative Path Delay Fault Simulator for Sequential Circuits. *Proceedings of IEEE International Test Conference*, October 1998.

[115] G. Parthasarathy and M. L. Bushnell. Towards Simultaneous Delay-Fault Built-In Self-Test and Partial-Scan Insertion. *Proceedings of 16th IEEE VLSI Test Symposium*, pages 210–217, April 1998.

[116] S. Patil and S. M. Reddy. A Test Generation System for Path Delay Faults. *Proceedings of IEEE International Conference on Computer Design*, pages 40–43, October 1989.

[117] A. Pierzynska and S. Pilarski. Pitfalls in Delay Fault Testing. *IEEE Transactions on Computer-Aided Design of Integrated Circuits and Systems*, 16(3):321–329, March 1997.

[118] I. Pomeranz and S. M. Reddy. Achieving Complete Delay Fault Testability by Extra Inputs. *Proceedings of IEEE International Test Conference*, pages 273–282, October 1991.

[119] I. Pomeranz and S. M. Reddy. At-Speed Delay Testing of Synchronous Sequential Circuits. *Proceedings of 29th Design Automation Conference*, pages 177–181, June 1992.

[120] I. Pomeranz and S. M. Reddy. The Multiple Observation Time Test Strategy. *IEEE Transactions on Computer-Aided Design of Integrated Circuits and Systems*, 41(5):627–637, May 1992.

[121] I. Pomeranz and S. M. Reddy. An Efficient Nonenumerative Method to Estimate the Path Delay Fault Coverage in Combinational Circuits. *IEEE Transactions on Computer-Aided Design of Integrated Circuits and Systems*, 13(2):240–250, February 1994.

[122] I. Pomeranz and S. M. Reddy. Design-for-Testability for Path Delay Faults for Large Combinational Circuits Using Test-Points. *Proceedings of 31st Design Automation Conference*, pages 358–364, June 1994.

[123] I. Pomeranz and S. M. Reddy. On Identifying Undetectable and Redundant Faults in Synchronous Sequential Circuits. *Proceedings of 12th IEEE VLSI Test Symposium*, pages 8–14, April 1994.

[124] I. Pomeranz and S. M. Reddy. SPADES-ACE: A Simulator for Path Delay Faults in Sequential Circuitswith Extensions to Arbitrary Clocking Schemes. *IEEE Transactions on Computer-Aided Design of Integrated Circuits and Systems*, 13(2):251–263, February 1994.

[125] I. Pomeranz and S. M. Reddy. On Synthesis-for-Testability of Combinational Logic Circuits. *Proceedings of 32nd Design Automation Conference*, pages 126–132, June 1995.

[126] A. K. Pramanick and S. M. Reddy. On the Detection of Delay Faults. *Proceedings of IEEE International Test Conference*, pages 845–856, September 1988.

[127] A. K. Pramanick and S. M. Reddy. On the Design of Path Delay Fault Testable Combinational Circuits. *Proceedings of 20th Fault Tolerant Computing Symposium*, pages 374–381, June 1990.

[128] A. K. Pramanick and S. M. Reddy. On the Fault Coverage of Gate Delay Fault Detecting Tests. *IEEE Transactions on Computer-Aided Design of Integrated Circuits and Systems*, 16(1):78–94, January 1997.

[129] S. M. Reddy, C. J. Lin, and S. Patil. An Automatic Test Pattern Generator for the Detection of Path Delay Faults. *Proceedings of IEEE/ACM International Conference on Computer-Aided Design*, pages 284–287, November 1987.

[130] K. Roy, K. De, J. A. Abraham, and S. Lusky. Synthesis of Delay Fault Testable Combinational Logic. *Proceedings of IEEE/ACM International Conference on Computer-Aided Design*, pages 418–421, November 1989.

[131] A. Saldanha, R. K. Brayton, and A. L. Sangiovanni-Vincentelli. Equivalence of Robust Delay-Fault and Single Stuck-Fault Test Generation. *Proceedings of 29th Design Automation Conference*, pages 173–176, June 1992.

[132] J. Savir. Skewed-Load Transition Test: Part I, Calculus. *Proceedings of IEEE International Test Conference*, pages 705–713, October 1992.

[133] J. Savir. Skewed-Load Transition Test: Part II, Coverage. *Proceedings of IEEE International Test Conference*, pages 714–722, October 1992.

[134] J. Savir. On Broad-Side Delay Testing. *Proceedings of 12th IEEE VLSI Test Symposium*, pages 284–290, April 1994.

[135] M. H. Schulz and F. Brglez. Accelerated Transition Fault Simulation. *Proceedings of 26th Design Automation Conference*, pages 237–243, June 1987.

[136] N. A. Sherwani. *Algorithms for VLSI Physical Design Automation*. Kluwer Academic Publishers, Boston, 1993.

[137] M. Sivaraman and A. Strojwas. Primitive Path Delay Fault Identification. *Proceedings of 10th International Conference on VLSI Design*, pages 95–100, January 1997.

[138] M. Sivaraman and A. Strojwas. *A Unified Approach for Timing Verification and Delay Fault Testing*. Kluwer Academic Publishers, Boston, 1998.

[139] G. L. Smith. Model for Delay Faults Based upon Paths. *Proceedings of IEEE International Test Conference*, pages 342–349, November 1985.

[140] U. Sparmann, D. Luxenburger, K.-T. Cheng, and S. M. Reddy. Fast Identification of Robust Dependent Path Delay Faults. *Proceedings of 32nd Design Automation Conference*, pages 119–125, June 1995.

[141] M. K. Srinivas, M. L. Bushnell, and V. D. Agrawal. Flags and Algebra for Sequential Circuit VNR Path Delay Fault Test Generation. *Proceedings of 10th International Conference on VLSI Design*, pages 88–94, January 1997.

[142] R. Tekumalla and P. R. Menon. Identifying Redundant Path Delay Faults in Sequential Circuits. *Proceedings of 9th International Conference on VLSI Design*, pages 406–411, January 1996.

[143] R. Tekumalla and P. R. Menon. Test Generation for Primitive Path Delay Faults in Combintional Circuits. *Proceedings of IEEE/ACM International Conference on Computer-Aided Design*, pages 636–641, November 1997.

[144] R. Tekumalla and P. R. Menon. On Primitive Fault Test Generation in Non-Scan Sequential Circuits. *Proceedings of IEEE/ACM International Conference on Computer-Aided Design*, November 1998.

[145] P. Uppaluri, U. Sparmann, and I. Pomeranz. On Minimizing the Number of Test Points Needed to Achieve Complete Robust Path

Delay Fault Testability. *Proceedings of 14th IEEE VLSI Test Symposium*, pages 288–295, May 1996.

[146] G. van Brakel, U. Glaser, H. G. Kerkhoff, and H. T. Vierhaus. Gate Delay Fault Test Generation for Non-Scan Circuits. *Proceedings of European Design and Test Conference*, pages 308–312, March 1995.

[147] H. T. Vierhaus, W. Meyer, and U. Glaser. CMOS Bridges and Resistive Transistor Faults: IDDQ versus Delay Effects. *Proceedings of IEEE International Test Conference*, pages 83–91, October 1993.

[148] R. L. Wadsack, J. M. Soden, R. K. Treece, M. R. Taylor, and C. F. Hawkins. CMOS IC Stuck-Open Fault Electrical Effects and Design Considerations. *Proceedings of IEEE International Test Conference*, pages 423–430, September 1989.

[149] K. D. Wagner and E. J. McCluskey. Effect of Supply Voltage on Circuit Propagation Delay and Test Application. *Proceedings of IEEE/ACM International Conference on Computer-Aided Design*, pages 42–44, November 1985.

[150] J. A. Waicukauski et al. Fault Simulation for Structured VLSI. *VLSI Design*, 6(12):20–32, December 1985.

[151] J. A. Waicukauski, E. Lindbloom, B. Rosen, and V. Iyengar. Transition Fault Simulation. *IEEE Design & Test of Computers*, 4(2):32–38, April 1987.

INDEX

I_{DDQ} testing, 33–35, 37, 38

absolute dominator gate, 141
algebraic factorization, 161
arrival time, 105
    earliest, 86, 88, 108
associated paths, 54
at-speed testing, 13, 78, 91
    on slow testers, 15, 19

Boolean factorization, 162
broad-side test, 10

cardinality of a co-sensitizing gate, 146
cardinality of a primitive fault, 65
co-sensitized paths, 65
co-sensitizing gate, 120–123, 147
    cardinality, 146
comparison block, 135
controlling value, 47
crosstalk, 3, 44, 169, 170, 172
    induced speedup/slowdown, 170

dag, 60
    leaf-dag, 60

deep submicron process, 1, 3, 5, 31, 44, 76, 169, 172
delay fault model, 23
    gate, 23, 27, 30, 38, 85, 134
    line, 23, 28, 30
    path, 2, 23, 28–30, 38, 157
    segment, 23, 29, 30
    transition, 23–25, 30, 33, 35, 37, 38, 44, 157, 162, 163, 171
delay-verifiable circuit, 167
directed acyclic graph, *see* dag
distributed defects, 1, 2, 23, 28, 169, 171
dominator gate, 141

enhanced scan flip-flops, 10, 12
enhanced scan testing, 10, 22, 159
enumerative path delay fault simulation, 78, 89

false path, 47, 54, 154, 155
fault activation, 12
fault initialization, 12
fault propagation, 12
five-valued logic system, 102, 103
functional irredundant path, 46, 56

functional justification, 10–12, 164
functional redundant path, 46, 56
functional sensitizable
    off-input, 53, 116
    path, 46, 47, 53, 54, 101, 113, 119
    tests, 101, 113–116
functional signal constraint, 71
functional unsensitizable
    off-input, 57
    path, 46
    segment, 58
functional vectors, 34, 44

gate delay fault, 23, 27, 30, 38, 85, 134
    simulation, 77, 85, 86
gross delay defect, 1, 33, 37, 38, 44
gross delay fault, 24

input sort, 61

latest stabilization time, 86, 88
line delay fault, 23, 28, 30

mandatory assignments, 108
merging gate, 65, 120, 123
multiple on-input, 64
multiple path delay fault, 46, 54, 64, 65, 99, 101, 102, 119, 130, 143
multiply-testable path, 63

noise faults, 169
non-controlling value, 47
non-enumerative path delay fault simulation, 78, 92, 93
non-primitive fault, 63
non-robust
    off-input, 51, 104, 105, 116
    testable path, 46, 47, 50, 51, 101, 104
    tests, 101, 104–107
    untestable path, 51

off-input, 47
    FS candidate, 116
    functional sensitizable, 53, 116
    functional unsensitizable, 57
    non-robust, 51, 104, 105, 116
    NR candidate, 109
    of a multiple path, 64
    partially static unsensitizable, 62
    robust, 49, 105, 116
on-input, 47
    of a multiple path, 64
one-hot coding, 163
optimistic update rule, 92

partial enhanced scan design, 12
partial scan design, 7, 9, 12, 13, 16, 18, 21, 131
partially static unsensitizable off-input, 62
path count reduction, 132, 143
path delay fault, 2, 23, 28–30, 38, 44, 45, 157
    classification, 45, 46
    cover, 66, 69
    coverage, 3, 36, 89, 93, 98, 131
    critical, 3, 31, 41, 42, 44, 132, 154, 171
    functional sensitizable, 113, 119
    multiple, 46, 54, 64, 65, 99, 101, 102, 119, 130, 143
    non-robust, 104, 171
    robust, 89, 102, 143, 171
    sensitization criterion, 46
    simulation, 77, 78, 88, 89, 91, 92
    single, 46, 65, 101, 102, 119, 123
    test generation, 101
    testability of a circuit, 131, 157
    untestable, 67
prime segment, 58
primitive fault, 63, 65, 119, 123
    cardinality, 65
    testability of a circuit, 144, 148
    tests, 102, 118
primitive path delay fault, *see* primitive fault
process variations, 1, 31, 44, 106, 170–172

rated speed testing, *see* at-speed testing
redundancy addition and removal, 137
robust
    dependent path, 59
    off-input, 49, 105, 116
    testability of a circuit, 143, 160
    testable path, 46–50, 101, 102, 143

INDEX 191

tests, 101, 102, 104, 157
untestable path, 50, 134, 157
robustness, 105, 107

scan shifting, 10, 12, 164
segment, 68, 69
segment delay fault, 23, 29, 30
    simulation, 78, 98
segment fault, 69, 70
    unexcitable, 72
    unpropagatable, 72
    untestable, 70, 71
sensitization criterion, 45, 46
    functional irredundant, 46, 56
    functional sensitizable, 46, 53
    functional unsensitizable, 46, 56, 57
    non-robust, 46, 50, 171
    robust, 46, 48, 171
    single-path, 46, 47
    static, 47, 64
    validatable non-robust, 46, 52
sequential path, 66, 68–71, 73, 75
sequential path delay fault cover, 68
Shannon's expansion theorem, 159, 160
single path delay fault, 46, 65, 101, 102, 119, 123
single-path sensitizable path, 46, 47
singly-testable dependent path, 63
singly-testable path, 62
six-valued logic system, 89, 91, 102
skewed-load test, 10
slack
    of a functional sensitizable test, 115
    of a non-robust test, 105
    of an off-input, 105
slow-fast-slow clock testing, 12, 66, 78, 91, 92
    on slow testers, 15–17
standard scan testing, 10, 11, 22, 159, 164

static sensitizable path, 47, 64
synthesis for delay-verifiable circuits, 166
synthesis for robust testability, 157, 159
    combinational design, 159
    enhanced scan design, 159
    standard scan design, 163
synthesis for validatable non-robust testability, 166

test application scheme, 7
    at-speed testing, 13, 15, 19, 26, 91
    enhanced scan testing, 10, 22, 159
    slow-fast-slow clock testing, 12, 15, 16, 66, 78, 91, 92
    standard scan testing, 10, 22, 159
test point insertion, 131, 133, 144, 146, 148
testing application scheme
    at-speed testing, 78
time-frame, 9
transistor
    input bridging fault, 37
    stuck-on fault, 37
    stuck-open fault, 24, 37
transition fault, 23–26, 30, 33, 35, 37, 38, 44, 157, 162, 163, 171
    simulation, 77, 78
    size, 25
true path, 47

unexcitable segment fault, 72
unpropagatable segment fault, 72
untestable path delay fault, 67
untestable segment fault, 71

validatable non-robust
    testable path, 46, 52, 101
    tests, 102, 112
variable speed testing, *see* slow-fast-slow clock testing